25 95

JEFFERSON COLLEGE

3 6021 00052 7998

D0882982

NO LONGER
PROPERTY OF
JEFFERSON
COLLEGE
LIBRARY

79-165

BF39
.M38
Vo2

The measurement of intrapersonal
space by grid technique.

DATE DUE

MAR 18			

Jefferson College Library
Hillsboro, Mo. 63050

Dimensions
of
Intrapersonal Space

The Measurement of Intrapersonal Space
by Grid Technique
Edited by Patrick Slater
Senior Lecturer, St. George's Hospital Medical School

Volume 2

Dimensions
of
Intrapersonal Space

Patrick Slater

with contributions from

Jane Chetwynd

J. C. Gower

J. P. N. Phillips

J. P. Watson

JOHN WILEY & SONS
London · New York · Sydney · Toronto

Jefferson College Library
Hillsboro, Mo. 63050

Copyright © 1977 by John Wiley & Sons Ltd.

All rights reserved.

No part of this book may be reproduced by any means, nor translated, nor transmitted into a machine language without the written permission of the publisher.

Library of Congress Cataloging in Publication Data:
Main entry under title:

The Measurement of intrapersonal space by grid technique.

   CONTENTS:
—v. 2. Slater, P., with contributions from Chetwynd, J., and others. Dimensions of intrapersonal space.
   Includes bibliographies and index.
   1. Psychometrics. 2. Least squares. 3. Personality assessment. I. Slater, Patrick. II. Title: Intrapersonal space by grid technique.

BF39.M38   155.2'83   76-8908

ISBN 0 471 99450 2

Typeset in IBM Journal by Preface Ltd., Salisbury, Wilts. and printed by The Pitman Press, Bath

PREFACE

It is a pleasure to me, and I hope it will be agreeable to you too, to trace the course that led me ultimately to the study of grid technique. Besides which, it will give me the opportunity of acknowledging my indebtedness to many people, some still alive and some long since dead, whose ideas have influenced me or who have helped me in other ways.

I have my chemistry master, Mr. Kitto, a very kindly sympathetic man, to thank for my introduction to psychology. When he discovered that I was keenly interested in ancient epics and sagas and primitive mythologies, he suggested that I should read Jung's *Psychology of the Unconscious*, which I did, if not from end to end. Jung introduced me to Schopenhauer and later, thanks to him, I was able to understand Kant, whose *Critique of Pure Reason* was one of the set books for Modern Greats at Oxford. Long afterwards I recognized his theory of knowledge as the original source from which personal construct theory and grid technique have descended.

The year when I came down, 1930, was not a good one for finding jobs, and my father, noting regretfully that my main academic qualification was in philosophy, or more precisely epistemology — not a very valuable asset in a job-hunt — suggested that until I could find employment I should occupy myself studying psychology as a student at the London School of Economics. This was a fortunate suggestion for there were no courses in psychology there at that time, and I had to attend University College where, as an external student, I was free to combine courses in psychology under Charles Spearman with courses in statistics under Egon Pearson. The modest knowledge of statistics which I acquired in this way gave me an advantage over other psychologists, who knew even less, and thereafter I was able to make a living combining research work in psychological measurement with clinical and assessment work.

For several years I applied factor analysis zealously to problems I encountered in research work under Alec Rodger at the National Institute of Industrial Psychology, but later, while working under Rex Knight in the Directorate of Selection of Personnel at the War Office, I began to lose faith in it. Euan Cooper-Willis, who was attached to the Adjutant General as a statistician with a roving commission to find out what the other statisticians in the War Office were doing, got to know me and asked me to explain factor analysis to him. I referred him to the classics of the time, Thomson's *Factorial Analysis of Human Abilities* and Burt's *Factors of the Mind*. Afterwards we had many discussions about the subject. He had studied Wittgenstein and argued cogently that factors could not be proved to be psychological entities but could only be regarded as classificatory terms applying to psychological variables which could be observed directly. At the same time I was learning that multiple regression equations and discriminant functions were much more efficient than factor analysis for solving the problems we were encountering in the allocation of personnel. They involved no reference to factors and often led to conclusions which were in conflict with factor-analytic expectations.

After the war I worked for a while in the Social Survey Division of the Central Office of Information with Leslie Wilkins, a man with a clear mind unconditioned by psychological indoctrination. One day he produced a table of correlations between a set of psychological variables which showed an unmistakeable general tendency towards negative correlation, and asked me what I thought of it. With my sceptical opinions about factor analysis I was fascinated. If a Thurstone analysis were applied to it, the first factor extracted would need to have imaginary loadings and then the residual variation would be greater than the total originally observed. This made a *reductio ad absurdum* of factor analysis.

From then on matrices of negative correlations became a hobby of mine. Nobody else seemed to be interested in them; I was on my own and could explore the whole subject one step at a time. I could make mistakes and retrieve them without embarrassment. For a long time I could only make intermittent studies of the subject, but after being appointed a full-time lecturer in psychological statistics I was able to investigate it systematically for a Ph.D. thesis.

First I had to assure myself that Wilkins' matrix was not unique. I searched for other examples and arrived at an experimental procedure that would always generate such matrices. Subjects were asked to evaluate a set of objects in terms of their personal preferences, using ranking, paired comparison or some intermediate form of multiple comparison. The data form a preference table, i.e. a two-way array with a row for each subject and a column for each object. When the preferences for the different objects are correlated, the resulting table of correlations is predominantly negative.

The need to extract an imaginary general factor is merely a paradox of Thurstone's method of analysis. Hotelling's method, principal component analysis, avoids the problem. In an experiment with m objects it accounts for the whole of the variation recorded in the preference table in terms of $m - 1$ dimensions. The result implies that while an m-dimensional reference system has been used for recording the variation there is one dimension where the experimental procedure has prevented it from actually occurring. Provided it has this effect any other procedure besides the one I had adopted will produce a matrix of predominantly negative correlations.

My results conflicted with the theory of specific factors which is incorporated in every method of factor analysis as distinct from principal component analysis. The theory had originated with Spearman, who noted a general tendency to positive correlation among the psychological tests in his experiments and explained it by supposing that they all measured a common factor, g, identified with general intelligence, but only to a limited extent — the rest of the variation they recorded was attributable to specific factors measured by each test independently of the others. With later extensions of factor analysis, evidence derived from the intercorrelations of homogeneous subsets of psychological variables has been used to support theories that other common factors exist which can be identified with particular aptitudes or personality traits; but the assumption has still been retained that after the common factors have been extracted from any such table, referring, say to m variables, there will still be residual variation in all the m dimensions, due to the independent specific factors. Thus the total number of factors postulated in any factor analysis exceeds the number of variables measured.

Principal component analysis, which originated with Karl Pearson (Egon's father), did not make any such assumption. It was proposed merely as a method for simplifying the records of a large number of correlated variables, not necessarily of psychological kinds, by reducing them to a possibly smaller number of independent measurements, ordered from largest to least according to the amount of variation they recorded. Uses for it have been found in other sciences but it has been rather neglected in psychology, where it is not considered suitable for testing the hypotheses proposed.

It becomes obvious when one thinks about it that variation recorded in terms of m coordinates cannot extend into more than m dimensions altogether and therefore that the assumption of specific factors as well as common factors must be fallacious. Objections of this kind are not new, but logical arguments unsupported by experimental evidence do not always carry much weight. Assumptions are frequently made for convenience without worrying overmuch about whether they are precisely true. And factor analysis has produced many useful results in mental measurement. It has been accepted by such eminent and influential psychologists as Burt, Thomson, Thurstone and Cattell – to mention only a few. It forms an important part of most courses on psychological statistics.

My experiments did more than simply disprove the theory of specific factors logically; they demonstrated that there is an important class of psychological data to which it cannot possibly be extended and, moreover, that the smallest dimension of variation recorded by a set of psychological variables can be brought within experimental control. These findings were liable to cause serious disturbance to people committed to maintaining factor-analytic theories.

Shortly after this research was finished, two Egyptian students, Sophia Magdi and Fawzeya Makhlouf, came to me for advice on analysing some grids they had collected. I found that the grids could be analysed by the methods I had developed for preference tables, and the students found that the results helped to make their data intelligible.

I for my part was fascinated by the potentialities of the technique. Though the data take the same form in a grid as in a preference table, their contents are profoundly different. The dispersion recorded in a preference table is macrocosmic: its coordinate system locates each informant at a particular point in a communal space of relative preference for different objects. The dispersion in a grid is microcosmic: its coordinate system records the distinctions one informant makes between certain things as they appear to him on a particular occasion. Thus grid technique extends measurement to variation within the intrapersonal space of a single individual, and in doing so makes advances of many kinds possible in psychology. For instance, the mental changes occurring in a patient during psychotherapy can be measured and the effects of the treatment can thus be assessed independently of the clinical impression of the therapist.

People using grid technique at that time did not appreciate its potentialities fully. They were usually content to examine the correlations between the constructs. Seeing that a grid is a two-way table showing the evaluation of a set of elements in terms of a set of constructs and that most, if not all, of the variation it records is due to the interactions between the two sets of functions, it is clear that examining the relationships between the constructs and ignoring those involving the elements must overlook an important part of the evidence. It also involves making the questionable assumption that the relationships between the constructs would remain the same if the set of elements was replaced by another one. On the other hand, sophisticated mathematical analyses would be needed to examine the relationships between all the functions of both kinds. They would be extremely laborious and tedious, and possibly inaccurate, unless carried out by a computer, which could do all that was required in next to no time.

Interest in personal construct theory and grid technique was increasing rapidly at that time (1960) and computers were also coming into general use; it seemed very likely that if the potentialities of grids as measuring devices were understood more completely and programs for analysing the data they yielded were accessible there would soon be many people eager to use them. I took a course in programming and started to develop programs specially for the purpose. Enquiries began to come in from many parts of the U.K.

These activities did not meet with approval where I worked. I was faced with overt indifference and covert obstruction. I ceased to be required to give courses of lectures and was refused financial assistance towards my computing expenses. During this difficult period I was particularly grateful for the lively interest my brother Eliot showed in my work, his steadfast moral support and his frequent practical advice. I was thankful, too, for the great tradition of academic freedom which protects people with unorthodox opinions in our universities, and was encouraged by the steadily increasing number of outsiders who came to me for advice and help. It was very satisfying to find that the work I was doing was so widely appreciated elsewhere.

Eventually there seemed to be a good case for applying for a grant to support the work. The Medical Research Council awarded me one in 1964 to provide part-time assistance for clerical work and processing data. The object of the research was twofold: to develop programs for analysing grids and to provide a computing service available to qualified psychiatrists and psychologists anywhere in the U.K. who were using grid technique in clinical work or academic research work. I considered it essential to combine the two sides of the work in order to take account of the needs of users in developing the programs.

The grant has been maintained up to this year, and increments have been approved from time to time as the amount of work has increased. Programs have been developed for comparing grids in various ways, but the main demand (amounting to about 80 per cent. of the total) is still for analysing grids individually. The program for the purpose has undergone many modifications. A version known as INGRID 67 has been superseded; the current version, described here in Chapters 7 to 9, was introduced in 1972 and accordingly is called INGRID 72.

My senior lectureship was allowed to lapse in 1970 and I was forced into premature retirement, but the M.R.C. allowed me a salary to continue with my work and added a salary for a research assistant in the initial grade of lecturer. I was fortunate to find Jane Chetwynd to join me with this appointment; she is a charming young lady who has relieved me of all the most troublesome part of the work, and contrived at the same time to get a Ph.D. and produce a long list of publications. Her loyalty has equalled her efficiency.

Thanks to Professor A. H. Crisp, who adopted grid technique as an aid to psychiatry as long ago as 1964, St. George's Hospital Medical School became the host institution for the project in 1973, and with the termination of the M.R.C. grant this year the School has continued to support and extend the work. The most important problem to consider next is how to adapt grid technique to measure disagreements between people who construe the same problem in different terms.

I owe thanks to many other people who have collaborated in the research at various times and to others who have helped me with their interest and encouragement, and apologize for not mentioning each of them personally.

Patrick Slater
4th February, 1976

CONTENTS

PART I

OBSERVATION

1

INTRAPERSONAL SPACE

1.1 Macrocosm and microcosm

There is no need of any exceptional insight to realize that everyone constructs an inner world for use as a plan of the outer world confronting him. He needs it to organize and understand his experiences and to direct his actions. Within it appear people he has met, things that have attracted his interest or aroused his fears, places he has visited or been told about — in sum, the whole extent and content of space, the past, the present and the future as far as he can visualize them. Moreover, many entities with little or no resemblance to anything in the outer world are probably included in it — the gods, saints and heroes, fairies, demons and monsters, the Utopias and El Dorados, the Heaven and Hell — rumours of which reverberate in his imagination.

Ideally the human cognitive system should be capacious enough to comprehend the entire universe and serve as a depository for all the ideas ever entertained about it. Its powers should extend even further, to take all sorts of possibilities and impossibilities into consideration as well as the actualities, to visualize previously unimagined things and to facilitate discovery and invention.

Leibnitz advanced the theory that every monad is a microcosm, a miniature replica complete in every detail of the macrocosm that contains it. He did not allow for individual differences and limitations or for spontaneous activity. Jung (1957) traced the remoter origins of the theory and enriched it with prolific investigations of the contents of human minds in abnormal as well as normal states. 'The psyche is the greatest of all cosmic wonders and the *sine qua non* of the world as an object,' he wrote; 'every science is a function of the psyche and all knowledge is rooted in it.' Neumann (1955 and cf. 1954) similarly regards the development of consciousness as the central factor in human history as a whole.

In characteristically vivid imagery Blake (1925) extended the idea to other living creatures:

When I came home, on the abyss of the five senses, where a flat-sided steep frowns over the present world, I saw a mighty devil folded in black clouds hovering on the sides of the rock; with corroding fires he wrote the following sentence now perceived by the minds of men and read by them on earth:

How do you know but every bird
    that cuts the airy way
Is an immense world of delight
    closed by your senses five?

(*Marriage of Heaven and Hell*)

Certainly the private universes of different individuals have a common structure, similar potentialities and limitations, and many similarities in content. The sensory system which receives the data is common to all humanity — though occasionally defective. So, too, is the cognitive system which organizes them and makes abstract reasoning possible in logic, mathematics and scientific method; otherwise arguments as abstract as those in Chapter 5 and some later parts of this book could not possibly carry any conviction. Moreover, a fund of common knowledge has been accumulating for millennia — common at least in the sense that its circulation is not effectively restricted. The parts officially considered most necessary are transmitted from one generation to the next by educational courses extending over the most receptive period of life. Other parts are fairly accessible in libraries and other sources of reference. An enormous amount of capital is invested in mass media for circulating ephemeral news and keeping public opinion informed — or formed in some other way. Common systems of belief and standards of conduct are maintained in particular cultures during certain eras, and deviations from them are treated as sacrilege or lunacy, or at least liable to some form of ostracism. Even in the most tolerant of communities the pressure to conform is powerful, insidious and unremitting.

Still some diversity of thought persists. Societies are bound to tolerate a certain amount and some may even encourage it in certain directions: the party in power may permit an organized opposition and grant it financial support, may leave the means of communication uncensored, may provide opportunities for private enterprise and may recognize academic freedom, regarding such liberties as salutary on the whole.

And there are innumerable respects in which each microcosm is unique. It belongs to a particular person. He has access to private sources of data in constructing it and is concerned in doing so with satisfying his personal needs or interests or avoiding their frustration.. There are vast areas in it distorted by prejudice and egocentricity or hidden away in the shadows of ignorance, forgetfulness, uncertainty and neglect. The elements are linked by a system of private relationships — feelings of affection, loyalty, desire, indifference, hostility, fear, and so on. It is hardly possible for anyone's microcosm to be more than an incomplete, distorted replica of the macrocosm to which it is intended to refer, since it is bound to need continual modification to accommodate the continuous input of fresh data.

The self with which a private universe is identified appears within that universe as a multifaceted self-image. Sometimes the realization dawns that it, too, must be incomplete and distorted — possibly on some trivial occasion such as brushing one's hair, looking at a reflection that shows the left and right sides correctly instead of reversed in a mirror, or listening to that affected or vulgar, shrill or booming voice on the tape-recorder repeating the feeble remarks that seemed so witty when one made them a moment before. Or perhaps the occasion may be some distinctly more embarrassing moment of self-revelation and disillusionment. Even then the self that appears is just another unexpected facet of the self-image, not a complete disclosure of the true self. Myself as I realize I am now is not likely to be much nearer the final version of the truth about me than myself as I used to think I was, or myself as my wife obviously sees me, etc. Who knows what the Day or Judgement will reveal?

However much one may wish to do so, one cannot transcend the boundaries of one's own inner world to attain the absolute truth about oneself or anything else. Truth is an ideal to be pursued. In the traditional saying that proclaims the creed of the scientist, *Magna est veritas et praevalebit*, the significance of the future tense is vital. Courts of law prescribe procedures in the endeavour to ensure that the evidence presented shall be the truth, the whole truth and nothing but the truth. The young scientist is trained to cultivate accuracy and objectivity in observation. Circulars arrive monthly and even weekly with information about new devices for increasing the accuracy or extending the range of scientific observations. Yet miscarriages of justice still occur and scientific experiments lead to dubious conclusions. The living truth flutters like a butterfly, tantalizingly away from the pursuer intent on pinning it down and preserving it.

The finest arenas for the rigorous exercise of accuracy and objectivity in observation are to be found in the physical sciences. Psychology has made many conscientious attempts to emulate them. Watson (1913, 1929, 1931) proposed that the subject should be restricted to the study of overt behaviour. He defined the subject matter of human psychology as the total behaviour of man from infancy to death. The same general types of method should be used for observing it as for the phenomena of all other natural sciences, e.g. chenistry, physics, physiology or biology. Subjective reports on mental states were not admissible evidence from his point of view (1929):

> So far in his objective study of man no behaviourist has observed anything that he can call consciousness, sensation, perception, imagery or will. Not finding these so-called mental processes in his observations he has reached the conclusion that all such terms can be dropped out of the description of man's activity. All behaviouristic observations can apparently be presented in the form of stimulus and response. The simple schema used is $S \rightarrow R$. A behaviouristic problem is solved when both the stimulus and the response are known.

With the spread of behaviourism the introspective methods developed by Wundt (1911) and extended by Titchener (1912) and his other students fell out of favour in the departments of psychology in the universities, particularly of the United States.

Mental measurement is another line of investigation that can be followed without depending on introspective evidence. The object is to make precise comparisons between people and not to discover what goes on within an individual. From the first, experiments in this domain, such as the ones devised by Galton (1883) for measuring sensory discrimination, have offered the observer a small number of simple responses to choose from, including one the experimenter considers correct. He merely notes whether that is the response given and scores the total number correct. He does not undertake any interviewing in depth to discover the reasons underlying the informant's choices. He can obtain all the evidence he needs without infringing the requirements of accuracy and objectivity.

(When he comes to reporting the results of his experiment he may, perhaps, exceed these limits by interpreting the variable he has measured as some postulated trait for which no objective criterion is available. But this may be avoidable. If he goes to the trouble of relating his measuring scale to some well-defined criterion he can complete his whole undertaking without any unsubstantiated speculation.

Given an unambiguous operational definition, this variable will serve to compare one informant with another; a trait-name to label it is superfluous and may lead to more misunderstanding than enlightenment.)

Vague, muddled thinking and casual observation are no doubt better than absolute obscurantism. They may at least help in the discovery of possibilities that would otherwise remain ignored. But clarity, precision and generality in reasoning and communication are as precious to science as accurate impartial observation. Verbal expression, whether in common speech or scientific terms, is often inadequate for describing the finer distinctions and more complicated relationships that need to be considered, and some kind of high level language based on mathematics and geometry needs to be employed.

The disciplines in psychology which have made the widest use of methods of enumeration measurement and estimation are behaviourism and trait psychology. Their progress has been achieved without undertaking detailed studies of mental content and processes in individual cases.

When it has suffered a breakdown is when the mind of an individual may stand most in need of detailed study. Here the psychiatrists have their domain. The possibility of relieving functional disorders by recollection under hypnosis was discovered by Breuer in 1880—82. He reported his procedure in the case of Fraulein Anna O. as follows:

> Each individual symptom in this complicated case was taken separately in hand; all the occasions on which it appeared were described in reverse order, starting before the time when the patient became bedridden and going back to the event which led to its first appearance. When this had been described the symptom was permanently removed (Breuer, J, reported in Freud, S. 1955).

Freud developed psychoanalysis from this starting point. He adopted free association instead of hypnotism, introduced the use of dream material, etc., and set the style of broad generalization supported by descriptive and anecdotal evidence emulated since by most of his colleagues and successors.

Among psychiatrists those who practice medical or surgical treatment and consider that discovering psychogenetic origins for mental disorders is not essential to effecting a cure have shown more readiness to employ objective methods. Those who undertake to explore the minds of their patients in depth by analysis are rarely interested in the possibility of using measurement and are sometimes reluctant even to consider the idea.

One may imagine arguments that could be used to support such attitudes. Firstly, psychiatrists and psychologists follow a well-established division of labour in their work together: the psychologist is primarily responsible for assessment and the psychiatrist for treatment, and therefore the psychiatrist leaves the study of methods of measurement to his colleague. Again, clinical and experimental procedures are profoundly different. The experimental psychologist sets up a standard situation, confronts his subject with it and expects him to give full attention to it. The psychotherapist encourages the patient to relax and allow his mind to wander without conscious direction. He may expect switching to an experimenal situation to have a very distracting effect. Other arguments may be advanced for considering psychometric methods unsuitable for psychiatric data. McCully (1971), for instance, quotes the view that experimental methods based on conscious logic are not readily adaptable to materials from outside conscious processes; the subjective can only be understood and judged subjectively. Such views are quite widely held.

Whatever the explanation, the fact is that comparatively little attention has been devoted to the problem of developing psychometric methods for measuring subjective variation. It is one that should not be so generally neglected. To take an example, let us compare the analogous statements, one macrocosmic and one microcosmic: 'S is more like X than Y' and 'S sees himself as more like Y than X'. Evidently the two statements may both be true though they do not coincide. One cannot say, off-hand, that the first must be taken into account and the second can be neglected. On the contrary, under some circumstances the second may be the more important one; it may, for instance, determine whose side S will join when X and Y start fighting.

The experimental psychologist may feel quite competent to verify the former statement with impartial objectivity using an appropriate set of scales — that is to say, scales appropriate for measuring variation in the macrocosm. If he is to tackle the problem of verifying the latter statement with anything like the same degree of confidence, what he will need will evidently be a set of scales appropriate for measuring the variation in S's microcosm, which cannot be assumed to be exactly the same as anyone else's.

Exploring a private universe is an undertaking psychology cannot avoid and is not yet adequately equipped to attempt. Maps giving distances and directions are an important part of the equipment in physical exploration; measurements are needed for constructing them; and methods for obtaining the measurements have to be studied. Similarly, methods of measuring the extent and direction of distances between points of reference in a private universe are prerequisite for mapping it and exploring it systematically. Their development has only recently begun and has tended to be haphazard.

1.2 Variation between and within persons

Psychological tests of a macrocosmic nature may be used as aids in diagnosis. They compare a particular person with others by pinpointing him on a scale or set of scales, and lead to conclusions of the logical form that he resembles certain people and differs from others in certain respects to a certain extent. This may help in assigning him to a diagnostic category, that is to say, to a class of people about which a fair amount is known, including what kinds of treatment are likely to be beneficial. Thus the usefulness of the tests must depend largely on the care which has been taken in standardizing them.

For example, the results of an intelligence test given as a measurement on the I.Q. scale is useful because of what is known about the distribution of I.Q.s in the population. An I.Q. of 120 would place a man at a point approximately 1.3 s.d. above the average and indicates that he is above 90 per cent. of the population and below the remaining 10 per cent. Such a result, locating an individual at a certain point on one macrocosmic scale, may evidently be very useful as a basis for recommendations about his treatment or disposal.

Similarly the measurements of an individual on two macrocosmic scales will place him at a particular point in a two-dimensional coordinate system. For

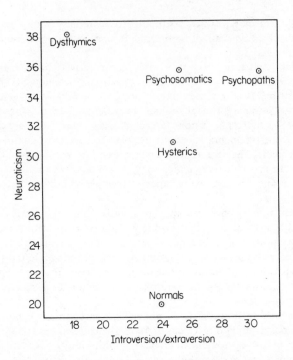

FIGURE 1.1 Mean scores of groups used in the standardization of the Maudsley Personality Inventory (Eysenck, 1959)

example, the Maudsley Personality Inventory (Eysenck, 1957) provides measurements on two scales, N 'neuroticism' and I/E 'intraversion/extraversion'. Figure 1.1 shows the coordinate system they form. An individual's measurements place him at a particular point on it. Say he is above average on the N scale and at the extraverted end of the I/E scale; this pair of measurements will place him at a point relatively near the mean for the psychopathic group. Or the combination of a high N score and a low I/E score will place him near the mean for the dysthymic group; and so on.

The tests given in a particular case may run up to any number, n. Then the scales will form an n-dimensional coordinate system and the individual's measurements will still place him at one particular point in it — not contariwise at n different points on one scale.

This is ignored when his measurements are used to make a profile for him. Then they may appear to contrast what he is like in some respects with what he is like in other respects, i.e. to compare him with himself and show within-individual variation. But, strictly speaking, an irregular profile does not provide a record of observed variation within the individual; it can only mediate inferences with limited degrees of probability about such eventualities.

Suppose, for example, his standardized scores on five attainment tests are

+1.3 s.d. on test P
+2.1 s.d. on test Q
+1.7 s.d. on test R
+0.1 s.d. on test S
+1.9 s.d. on test T

The profile in Figure 1.2 which represents them shows that though his score on test S is slightly above the average for the population it is well below his own scores on the other tests. One might surmise that if he is aware of this relative weakness he may feel that it cramps his style in some respect; he may be exposed to stress of some particular kind, resulting in a disturbed mental state or occupational maladjustment.

This is a microcosmic interpretation placed, with a limited degree of confidence, on observations which are themselves purely macrocosmic. It cannot be reached without comparing the observed S score, implicitly at least, with an estimation of what it should be in this case; and the estimate can only be derived from observations of the distribution of the test scores in the population. If the measurements could not be expressed in standard scores making use of norms for the population, they could not sustain the inferences placed on them at all.

The entire range of variation exhibited in the distribution is due to differences between persons. Each individual appears in it as a single point. The variation within the individual, the universe within the point, is not exposed to observation. Yet variation is certainly to be found within him when he is viewed as a microcosm, because it contains many elements distinguished from one another in many respects and degrees. Methods may be devised for measuring such variation.

Grid technique, with which this book is concerned, is one particularly interesting method. It provides a coordinate system which can be extended indefinitely and adapted specially to fit a particular individual. It does not require comparisons with any other cases. So the psychological demands it satisfies and those satisfied by

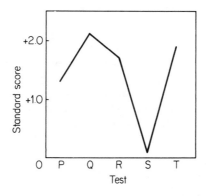

FIGURE 1.2   An illustrative profile

standardized tests have nothing necessarily in common. It deserves a place of its own in clinical work and may eventually prove as important there as standardized testing procedures; but it is not a substitute. The contributions in the companion volume, *Explorations*, also illustrate other applications that can be found for it.

1.3 Variation within an individual

Let us first examine some of the other ways in which variation within an individual can be shown to be

(a) open to observation and measurement,
(b) orderly and
(c) independent of experimental control.

We may then be more prepared to find that microcosmic variation had these properties too, or at least feel less inclined to doubt whether it exists or is anything else than macrocosmic variation all over again. These assertions are not altogether beyond the reach of controversy though plenty of evidence can be found to support them.

A simple and common experimental model providing measurements of variation within an individual is one where subjects, perhaps of several classes, undergo repeated trials, perhaps under different treatments, and a measurement is recorded for each subject at each trial. Then the variation observed within an individual is the amount found between the measurements in the set referring to him.

If an analysis of variance is made in such an experiment a general average is calculated of all the measurements recorded for every subject. Particular measurements naturally differ from it more or less. The total observed variation is obtained as the sum of squares of all the differences. Part is due to variation between the subjects and part to variation between the trials; and there is also a residual part, attributable to 'subject/trial interaction', for naturally each trial does not affect each subject in precisely the same way. The variation within the individuals includes the second and third parts; variation between them only accounts for the first.

The residual part is sometimes treated as negligible and labelled 'error variance'. For example, an intelligence test may be given on two occasions. Then (even though a different method of analysis is more often used) its test—retest reliability is measured in effect by comparing part one in the analysis of variance, i.e. the amount of variation between the subjects, with part three, the amount due to interaction. The postulate that a subject's intelligence does not vary from trial to trial forms part of the accepted notion of intelligence and if the amount of interaction is relatively large the test is condemned as unreliable.

This assumption *that subject/trial interaction is entirely erratic* is habitual in the whole field of cognitive testing. The first question asked about a test is 'What is its reliability?', and it is considered unfit for publication unless a large-scale experiment shows that its reliability is high.

Thus trait psychology shows a tendency to concentrate attention on the more stable aspects of personality, and perhaps to treat aspects which may be unstable as more stable than they necessarily are. Attempts to treat neuroticism as a constitutional trait are open to criticism in this respect. Neurotic disturbances occur in heterogeneous cases, can be precipitated in different ways, produce different symptoms and respond to different treatments. It should be possible to develop tests to measure variable states of mind and so, for instance, trace the changes in a patient's mental state from the onset of a nervous breakdown to the point of recovery from it. But ones which are suitable for such purposes may not do well in the usual test—retest experiments for assessing reliability. Accordingly they are likely to appear suspect to the trait psychologist and the idea of developing them may be found uncongenial.

In more elaborate experiments with several classes of subjects, several treatments and several trials for each treatment, analysis of variance can be used to extract amounts due to differences between

    classes
    subjects in the same class
    treatments
    trials for the same treatment

and to a whole series of interactions

    class/treatment
    subject/treatment within class
    class/trial within treatment
    subject/trial within treatment within class

Differences between persons only contribute the first two parts of the total variation — between classes and between subjects within class — all the rest of the variation is within persons.

So in such an experiment one kind of variance within persons can be compared with another kind, and significant differences may be found. It is possible to test whether the subjects reacted to the treatments uniformly or differently. The evidence will prove that their ways of responding to the same treatment are different and characterize them as individuals and not as members of a class, if the subject/treatment variance within class is signficantly greater than the subject/trial variance within treatment within class. Examples of experiments where such results have been obtained are too common to enumerate. In one (Gray, 1965) the subjects compared were untrained rats of the same sex and age from two pure-bred strains.

Or for a simple analogy we might record a child's weight accurately twice daily every day for a month; analyse the total variation in terms of continuous growth, regular daily fluctuation and accidental deviations; and test whether the first two components are significantly greater than the third. One child would be enough for such an investigation; we would not need to compare him with any others. But including others would enable us to test the significance of individual differences in growth rates and diurnal fluctuations. In geometrical terms the results for an individual from such an experiment are a set of points which are not scattered at random but form an orderly, unique array.

1.4 Coordinate systems for measuring microcosmic variation

As microcosms are images of the macrocosm, demographic experiments applicable to the macrocosm should generally provide models for idiographic experiments to measure variation within a microcosm.

A simple demographic experiment where two objective tests are given to a class of school children might be taken as a model and reproduced idiographically by asking the class teacher to express his opinions of the pupils in his own terms, using two rating scales improvised for the purpose. The role of the experimenter could remain impersonal in both experiments. The first would provide acceptable evidence of variation between the pupils in an objective test-space; the second would not. But it would provide evidence of variation within the microcosm of the teacher, revealing what he thinks of his pupils and quantifying the distinctions he draws between them.

The two sets of data would have the same form though they differ in content. They can both be recorded in the same layout: a two-way table listing the measurements of all the subjects on each of the measuring scales. By simply looking at the two tables without access to any further information it might not be possible to tell which came from which experiment.

Diagrams can be constructed in the same way to exhibit the data from both experiments. A surface can be marked out like Figure 1, with its horizontal and vertical axes for the two scales and each pupil's position on it marked at the point with the coordinates given by his two measurements. Then the dispersion in the idiographic figure would map a cross-section of part of the teacher's microcosm just as the demographic figure maps the dispersion of the pupils in the objective test-space.

The same mathematical methods can be used to analyse the data. Means, standard deviations and correlation coefficients can be calculated. The connection between the two rating scales in the teacher's personal system for assessing his pupils can be expressed in the same notation as used for describing the objective association between the two tests.

The number of scales can be increased. Objective tests suitable for classes of school children are in plentiful supply. Nor are teachers likely to exhaust their opinions of their pupils by rating them on two scales — most can think of a dozen or more ways of assessing them. Within reasonable limits, measurements of m cases on n scales can be obtained as readily with either procedure. Methods of multivariate analysis developed for demographic data can be adapted for idiographic purposes.

Just as a school teacher may assess his pupils a theatrical critic may assess the shows in town or an art critic the pictures in a gallery, a fashion critic a Paris collection, a television viewer the week's programmes, a car dealer motors, a cook vegetables, and so on. Each may be expected to have his own criteria for judging the things with which he is personally concerned. If as an experiment he assesses them on a numerical scale, the operation of judging the things according to the criteria will generate a set of commensurate numbers which may be listed in a two-way table. A column may be assigned to each thing and a row to each criterion, and the cell in row i, column j will contain the assessment of j, one of the things, in terms of i, one of the criteria.

In Kelly's (1955) terms the operation is *construing*, the things construed are *elements*, the criteria for judging them are *constructs* and the table recording the results of the operation is a *grid*. These convenient terms will be retained. Briefly, then, it is to be understood here that the general form of a grid is a table with n rows and m columns, recording one informant's assessments of m elements in terms of n constructs. All the entries in it are known, commensurate numbers describing variation within his microcosm.

The manner in which they are used is what distinguishes the elements from the constructs. The terms for them can be interchanged, sometimes if not always. For instance, in saying 'Peter is like Paul' an informant applies the construct 'like Paul' to the element 'Peter', and the interviewer could extend the construct to other elements by naming other people and asking whether they are more like Paul or less. If the statement were 'Paul is like Peter' the functions of the two terms would be exchanges and the interviewer would proceed in a different direction. Constructs, we may say, function as operators and elements as operands. Though terms may change functions, it is essential for a grid to contain two sets of terms, one for each function, and a rating scale for defining how far the operators apply to the operands.

The contents of a grid are bound to be restricted. It can amount to no more than a single exposure — a snap of a small part of a private universe. Even a carefully planned series of experiments with one informant may reveal his universe no better than his social life might be revealed in an album of snaps. Other methods of gaining insight into it have long been in use and will no doubt continue to be needed.

However, the potentialities of grid technique should not be underestimated. Their distinction is that they extend the reach of measurement to intrapersonal space and enable the relative amounts of variation in different directions within it to be compared. How much information can be extracted from one grid by efficient analysis is not yet fully appreciated. Moreover, new techniques for obtaining grids are being tried out and new methods for comparing them in sets are being formulated. Developments are thus proceeding at an accelerating rate; the scope for them is still enormous and the limits are far out of sight.

2

GRID TECHNIQUE IN HISTORICAL PERSPECTIVE

2.1 Introduction

Any method of interviewing may be described as a grid technique if it can be used to obtain enough data from one informant to complete a grid. Methods which are not intended for that purpose and ones which do not provide enough data when carried out in accordance with their original directions are yet potentially a grid technique if they can be pursued far enough and recorded in sufficient detail to provide all the data needed to locate the elements and the constructs in the space under observation.

It may be felt that there is some impropriety in examining such methods without considering the purposes for which they were originally intended. The authors who have proposed them generally consider their purposes paramount and their methods only of subordinate interest. But when the method they choose for obtaining the measurements they need turns out to be essentially the same whatever the theory it is harnessed to, its full capabilities must appear worthy of attention. We should look, for a change, at the horse instead of the goods displayed on the cart it happens to be pulling at the time.

2.2 Moreno's sociometric test

This is one of the earliest reported procedures (Moreno, 1934) adaptable for use as a grid technique. It is intended particularly for studying interpersonal relationships in the spontaneous formation of social groups, and is applied collectively to the members of an organized group, preferably a fairly well-defined one such as the pupils in a school class or the inmates of one of those institutions sometimes rather euphemistically called 'homes'. For example, the pupils might be told: 'You are now given the opportunity to choose the boy or girl you would like to sit on each side of you next term. Look around and make up your mind. Write down first the one you would like best, and then the one you would like second best. Remember that next term the friends you choose now may be sitting beside you.'

Moreno emphasizes that the procedure should not require the informants to assume a passive submissive attitude towards the instructions they are given, but should stimulate their interest by referring to some practical end they wish to realize. Their feelings towards one another must be defined in terms of a specific criterion, such as living together, working together or sexual attraction. Simply asking them whom they like or dislike in their community irrespective of any criterion is not sufficient. Their answers should be given in the expectation that they will have definite consequences. They should be taken into partnership and become sufficiently interested in the test to reveal their spontaneous attitudes, thoughts and motivations with reference to the other people concerned. For them the test should not be a test at all, but an opportunity to take an active part in matters affecting their personal situations.

Nor for that matter is it a test at all for the orthodox psychometric psychologist. It would be far better described as an interviewing procedure. None of the methods recommended for constructing tests — item analysis, standardization, validation, etc. — are applicable to it. On the other hand, it has the compensating advantage of being adapted to the situation in which an informant finds himself at the time and the feelings he experiences in consequence. The flexibility of grid techniques is its most conspicuous advantage over other psychometric methods. As different variants of it are examined, the range of situations to which it can be adapted will come into clearer view.

The sociometric test as described is only an embryonic form of grid technique. It cannot generate a complete grid without some extensions or modifications. Moreover, it does not measure variation in an unambiguously defined space, but confuses variation between persons with variation within.

The additional information needed from a single informant to complete a grid could all be obtained by simple extension of the procedure outlined by Moreno. Instead of just one criterion likely to influence the spontaneous formation of a group, a series of criteria should be considered. For instance, the question Moreno supplied for the classroom enquiry might be supplemented by others such as:

> Who would you most like to have
> > on your side in a game of football?
> > to give you help with your prep?
> > to go camping with? etc.

It would also be better if each pupil could be interviewed separately and encouraged to define any other criteria that affect his personal preferences.

He should next be asked to apply the criteria to all his classmates, including those he would rather keep out of any of the groups to which he belongs as well as those who would be welcome. He could rank them or grade them on any convenient scale. His responses would then constitute a grid, with the criteria as constructs and his class mates as elements, indirectly describing the relationships between them as they appear to him personally.

The distance between two points can be measured in a space where they are both given a definite location. A grid records the location of the elements and the constructs in the informant's intrapersonal space. Consequently, one obtained in the way described will measure the distance between his classmates in the informant's space.

Grids obtained from all the pupils, A, B, C, . . . , can be used to measure the distances between A and B in the intrapersonal spaces of C, D, E, . . . ; between A and C in the intrapersonal space of B, D, E, . . . , etc. But the distance between A and his classmates cannot be measured in his own space because his grid does not locate him in it. He remains outside.

There is no absolute necessity for this. The difficulty could be overcome by using constructs of a different kind or by modifying the procedure in some other way. It only arises because the constructs proposed for the sociometric test do not include the informant within their range of convenience. As Kelly would say: a pupil can hardly be asked how much he would like to have himself sitting beside himself in class, for instance.

Obtaining separate grids from all the members of the group and measuring distances in each intrapersonal space might be considered an unnecessary elaboration of the experiment. The sociometric problem concerns the interactions between people brought together by external forces into an assembly where common and conflicting interests unite some and separate others — in short, it concerns the variation between the members of the group, not the variation within their private systems. A single measure of every distance, incorporating the opinions of every member, should be enough, it might be said, to indicate how they coalesce or separate out.

If measurements of this kind are required, an alternative procedure would be more suitable for the purpose. A possibility would be to treat the whole group as a corporate person with its own private universe — a universe of interests specific to it as a whole and affecting its members in different ways. Each member might be asked to define his attitude to each of these interests on a scale ranging from strongly pro to strongly anti. The complete set of data so obtained could be arrayed in a single grid where the interests function as elements and the members as constructs. Their locations would be defined in the corporate space and the distances between them could be measured.

Now the members have been introduced as authors of grids referring to their own intrapersonal spaces, as elements in the grids recorded by other members and finally as constructs in a collective grid. Definite measurements of distances in the different spaces can be obtained. It is for the experimenter to decide which are the most important ones for his purposes.

2.3 Stephenson's Q-technique

Arrays which take the form of grids are obtainable by Q-technique, and some have been reported by Stephenson personally.

He began developing the technique in the 'thirties (Stephenson, 1935) with the proposal to use the subjects' scores on test to correlate the subjects instead of the tests. To distinguish between the two procedures he referred to them as Q-techniques and R-technique respectively. Both could start with the same array of data, say A, a table with s rows and t columns recording the scores of s subjects on t tests. Correlating the tests would derive a table R with t rows and columns, where the typical entry, in row i column j, is the correlation between tests i and j. This practice had been almost universal for many years (Thomson, 1948, mentions some exceptions). Q-techniques would derive a table Q with s rows and columns, where the entry in row i column j is the correlation between subjects i and j.

The idea come under severe criticism from two leading British exponents of the traditional technique, Burt (1940) and Thomson (1948). In spite of this discouragement Stephenson continued to elaborate Q-technique after his appointment as professor of psychology in the University of Chicago. His authoritative exposition of it, which was published in 1953, describes experimental methods for obtaining data to be used for correlating persons and also other methods of analysis as well as correlation and factor analysis which can be applied to such data. Examples are included.

Adopting Q-techniques leads to a different way of designing experiments. The two correlation tables Q and R obtained from the same array of data, A, are generally of different sizes. Q is larger than R when there are more subjects than tests, and vice versa. If the results from both techniques converge, obviously R is preferable for simplicity when s exceeds t and Q when t exceeds s. Indeed, whether convergence is conceded or not, those in favour of Q-technique must favour giving many tests or putting many questions to a few subjects — it would be a cumbersome way of treating data from experiments following the traditional designs for R-technique. Accordingly, Stephenson's researches advanced towards more intensive studies of fewer cases. Finally he arrived at methods for obtaining a whole two-way array of data from one informant; and he describes Q-technique as 'a methodology for the single case'.

Another reason for designing experiments differently was that a problem had been encountered in applying Q-technique to scores on tests of the kinds used in experiments with R-technique. Such scores are not generally commensurate: every test has a different mean and standard deviation. Some preliminary processing which makes scores on the different tests commensurate is needed before they can be correlated. Stephenson, Burt and Thomson had differed about the corresponding processing needed for correlating the subjects. To avoid the problem, Stephenson replaced tests by single items, such as statements from questionnaires, and instructed his subjects to rate them with a rather sophisticated rating scale called a Q-sort. This assigns them all commensurate values. One of his suggestions was to use the forty-eight photographs from the Szondi test as items (c.f. Szondi, Moser and Webb, 1959). After they had been shuffled, an informant might be asked to characterize them in various ways: rating them on occasion A from the ones he liked most to least, on B from most godfearing to least, C handsomest, D healthiest, E oldest, etc. Stephenson proposed that the scores assigned to the items on scales A, B, C, . . . should be correlated and factored, all for the one informant. In our terms the data from such an experiment could be presented as a grid where the elements are the photographs and the constructs are the characteristics in terms of which they are evaluated.

In another study an informant provided fifteen self-descriptions in terms of sixty self-referent statements, such as

> I rarely become very excited or thrilled
> I often become entirely absorbed in thinking about myself
> I feel pleasantly exhilarated when all eyes are on me
> I sometimes feel as though I should run out of the room I am in, scream or burst into tears

etc. He assessed how well these statements applied to him

> as I am at this time
> as I would like to be ideally
> as I was between the ages of nine and sixteen (two sorts)

as my friends regard me
as Mother saw me
as Father saw me
as Mother wanted me

etc. In the grid so generated the statements would be the elements and the self-descriptions or self-images the constructs. Stephenson reports the results of a centroid analysis at length.

In the first experiment the informant was to make his ratings on a nine-point scale with a prescribed frequency distribution (Table 2.1). The two photographs rated highest were to be given a score of 8, the next four a score of 7, and so on down to a score of 0 for the two lowest. Such prescribed distributions are a distinctive feature of Stephenson's Q-technique. He gives many examples of spreading different numbers of items over ranges with different limits — mostly exceeding the nine-point scale.

The most explicit description he gives of the procedure is with reference to a lady asked to sort self-referent statements:

> The statements are typed on cards, one statement to a card, and the subject first reads them through in order to grasp their import. They are then shuffled, and she proceeds to the Q-sort. Usually the cards are first divided roughly into three piles by the subject, one for those that characterize her positively, one for those that do not do so, with the doubtful or neutral ones in between. The piles are then teased apart until the required frequency is reached.

Other points he mentions are:

> All the statements have to be compared with one another, and judgements must be made about each statement in the context of all the others and the conditions of instruction ... the operator sorts them into ten or more classes on a quasi-normal frequency basis ... we follow the practice of using a much flattened symmetrical distribution of scores for all Q-sorts ... everyone has to follow the same distribution.

Stephenson does not provide adequate justification for his insistence that the same form of distribution should be maintained in every experiment and that all variables in a particular experiment should have precisely the same distribution, but it is intelligible as a means of securing commensurate data throughout. Most experimenters would consider it unnecessarily rigid and artificial, and it seems to be rather inconsistent with his main objective of making each informant the subject of an individual study and dispensing with standardized test material.

Table 2.1

	Most liked								Least liked
Score	8	7	6	5	4	3	2	1	0
Frequency	2	4	5	8	10	8	5	4	2

2.4 Osgood's semantic differential

A semantic differential is a form of grid. It records an informant's evaluations of a set of concepts equivalent to elements in terms of a set of scales equivalent to constructs, and can be used to explore the informant's intrapersonal space. That, however, is not the purpose for which it was designed.

Human beings are naturally gregarious and form groups of all sizes from the nuclear family to international alliances and syndicates. The members of such groups are united by common interests and their solidarity may often be reinforced by opposition to the interests of other groups. It may be said that members of any such group share a common meaning-space and communicate with one another by using terms which have recognized locations in it. The common meaning-spaces or semantic-spaces are conceivable for groups of all sizes up to and including humanity as a whole.

The semantic differential is designed as an instrument for measuring variation within such spaces. It is meant to be given to subjects in groups and the instructions for administering it are quite explicit, but it is not a rigidly constructed test. It is a flexible technique which can be adapted to the needs of different projects by varying the concepts and the scales.

Osgood (c.f. Osgood, Suci and Tannenbaum, 1957) is deeply committed to the view that three dimensions are to be found in the meaning-space of almost every group. The major dimension is evaluative, including scales such as

    good — bad
    beautiful — ugly
    fragrant — foul
    sweet — sour
    clean — dirty
    pleasant — unpleasant

The second refers to potency, including

    strong — weak
    large — small
    heavy — light

and the third to activity, including

    fast — slow
    active — passive
    sharp — dull

In some groups the dimensions of activity and potency tend to merge, but the evaluative dimension is generally the most conspicuous.

Every scale is defined by contrasting two terms as its opposite poles, and is graduated at seven points: *sweet—sour* for instance, would be graduated very sweet, quite sweet, slightly sweet, neutral, slightly sour, quite sour, very sour.

Osgood gives a long list of scales which have been validated experimentally and proved suitable for measuring variation in his three main dimensions. He advises people who use the technique to include at least three scales of each kind in their differential, to provide the subject with 'a balanced space which he may actually use as he sees fit. If he makes more discriminative use of the evaluative factor relative along to others this will show up in his data (in an elongation of his space along this dimension) but he is not forced by the sample of scales to do so.' The experimenter may add new scales with no known relationship to evaluation, potency or activity; and the results will reveal their relationships, provided the three major dimensions are well defined.

The concepts should be diversified enough to cover the whole area of interest and should be familiar and unambiguous to the subjects. And the scales should be relevant to the concepts. For example, if a sample of the electorate were asked to rate the leaders of the political parties, *beautiful—ugly* might be discarded as an evaluative scale in favour of *sincere—insincere*; and scales such as *radical—conservative* might be introduced for the occasion. The results of such an experiment might be used to locate the politicians in the semantic space of the electorate and measure the distances between them, as well as to discover the relationship between the scales specially introduced and those used to establish the main dimensions of the space.

A large-scale investigation would presumably be preceded by a pilot study of the usual kind, with a small sample of subjects, possibly a structured one, from the population concerned. Extra concepts and scales could be tried out and some unstructured supplementary interviewing could be conducted to discover how they were being interpreted. Thus there would be consultations between the investigator and his subjects at this stage, and he would base his final choice of concepts and scales on the results. The sample of subjects for the main survey would be a different one, drawn at randon from the same population.

The differential is presented in the form of a questionnaire. Osgood has drafted typical instructions to appear on the front page explaining how it is to be completed. Each of the following pages is headed by one concept. The scales are listed beneath, with the contrasted terms at the opposite ends of a line marked off into seven sections thus

high: :   :   :   :   :   :   : low

The subject records his grading of the concept by marking the appropriate section.

Scales and concepts were alternated in the original prototype of the differential (Stagner and Osgood, 1946): a concept was presented on the left of each line with the scale for grading it to the right; the next line gave another concept and scale. The intervals between recurrences of both were maximized. This eliminates the 'halo' effect by preventing the subject comparing his gradings of a concept on different scales. But it makes the questionnaire more difficult to produce and to score, and makes filling it in more complicated and confusing for the subject. Negligible differences were found between the mean gradings of the concepts on the scales in an experiment where the two methods were compared (Osgood, Suci and Tannenbaum, 1957). So the simpler method was recommended. Scales referring to different dimensions can be juxtaposed in it and their polarities alternated to prevent the formation of response sets.

2.5. Kelly's theory of personal constructs

Grid technique and Kelly's theory are not indispensable to each other. The theory can be formulated without reference to the technique and put into practice without applying it. According to Hinkle (1970), Kelly proposed omitting the chapters on it from the revised edition of *The Psychology of Personal Constructs* which he had undertaken before he died; and the abridged edition, *A Theory of Personality* (1963), includes no references to it. Conversely, as the procedures already reviewed exemplify, grids are obtainable without depending on the theory. Osgood in particular laboured, not altogether successfully perhaps, to reconcile his technique with behaviourist doctrines which Kelly treated as anathema.

There is, however, an intimate connection between the two. The theory explains the technique. It shifts the focus of interest from interpersonal to intrapersonal variation, and in doing so it sets aside the rules for constructing objective psychometric scales enunciated by the majestic succession of psychologists from Galton to Guilford. In describing the technique Kelly gives general directions on preparing a grid specially for a single individual, as an instrument for measuring how things appear to him personally; and in deriving the technique directly from the theory he provides a rationale and a terminology for the whole process of constructing the grid, administering it, and interpreting and applying the results. His theory may be acceptable pragmatically for this purpose by people who do not find it entirely satisfying or who even regard it as objectionable in some respects.

George A. Kelly graduated in engineering with psychology as a subsidiary subject at the time when the boom of the 'twenties in the United States was collapsing into the depression of the early 'thirties. He found it impossible to obtain employment as an engineer and went to practise psychology in an impoverished, inaccessible area on the border of the dust-bowl. There he had to do the best he could to diagnose and treat the peculiar psychological problems of his patients by himself, without the advice or assistance of any medically qualified psychiatrist or other psychologist within reach for consultation. He found little help in the popular psychologies of the period, behaviourism and psychoanalysis. Through trying many experimental approaches he gradually developed a homespun psychology which he expounded in his major work, *The Psychology of Personal Constructs* (1955), and elaborated in a very large number of other publications after he entered the academic world.

From what sources Kelly assimilated his opinions is a question that can only be answered speculatively; there are too few references in his publications for their origins to be discovered. He does not support his theory by appeals to authority but presents it tentatively as his own. Others may accept it or reject it as they please; he simply invites them to try it out. His criticism of others, e.g. his attacks on behaviourism, are less effective, as Holland (1970) observes, because they likewise lack references.

Commentators are inclined to consider that philosophy contributed more than psychology. His notion that nothing can be known except through a personal construct system which may err and can be changed goes back to Kant and earlier idealists, as Oliver (1970) notes; and his formal exposition of his theory in terms of a primary postulate and eleven corollaries recalls Spinoza. Hinkle (1970) compares him with Heraclitus, Wittgenstein and Polenyi, and Holland (1970) refers to many other philosophers. His theory might indeed be classed as an epistemology rather than a psychology.

Construing, which he regarded as the universal or at least the typical form of thought, is the recognition of a contrast between two sets of things. A man construes his acquaintances, for instance, when he forms the opinion that some are friendly and others hostile to him. The construct *friendly/hostile* would be useless to him if it did not provide any distinction between people he has to mix with. Even though only one of its opposite poles is defined a construct is implicitly bipolar. The term used to define it explicitly is described in that case as its emergent pole and the contrast left undefined as its latent pole.

Everyone applies his constructs in his own way. One man's friends and foes need not be the same as another's, nor the political affiliations or social activities that attract or repel him. Even constructs which appear unequivocal, e.g. *British/foreign*, may be applied idiosyncratically by different people. No one has yet been wise enough to propound a universal system, acceptable to everyone and applicable to everything.

Kelly's fundamental postulate is that a man's behaviour is directed by the way in which he anticipates events. He does not simply let them push him around; he attempts to predict and control them. He develops his construct system to make sense of his environment and chart the course of his future behaviour. In the perspective of history man may be seen as an incipient scientist. Any alert person experiments with his constructs, adopting them tentatively and testing them at every opportunity. Each day's experience may confirm some parts of his construct system and call for revision or outright abandonment of others.

At any given time his construct system consists of a set of miniature systems with limited ranges of convenience to deal with different aspects of his environment. One, for instance, may apply to the people he knows, another to openings for jobs he is considering, another to his choice of breakfast cereals, and so on. Every construct has a limited range of convenience where it applies most effectively.

One part of his construct system, however, tends to become relatively sacrosanct — his core structure, the set of constructs which collectively defines the role he sees himself called upon to play in life and helps him to maintain his sense of self-identify under the bludgeonings of chance. He tends to adopt courses of action which corroborate it and avoid behaviour which would invalidate it. Interference with it is liable to generate mental stress. He may experience feelings of anxiety when confronted with events outside its range of convenience or guilt feelings when forced to play a role inconsistent with it. Thus Kelly expresses mental aberrations as well as normal mental processes in terms of the functioning of personal construct systems.

Therapy may involve inducing a patient to accommodate himself to necessary modifications in his construct system while reducing the threat of disruption to his core structure. He may be encouraged, for instance, to try adopting a different role experimentally for a limited period — one suggested by the therapist with a view to helping him discover a possible escape from his predicament. Grid technique may assist the process by defining his problem for him in terms that he can recognize and accept, and by revealing ways of reconstruing it.

Evidently Kelly's psychology was clinically orientated in its origin or, as he would say, has psychotherapy as its focus of convenience. But it can be extended indefinitely. He found that it could be applied to the problems his students brought him for discussion in connection with their researches as well as to the personal problems brought by his patients. Any term of discourse whatever, thing or theory, considered by an individual can be treated as a point of intersection of certain constructs in his system as a whole. Thus in its full extent the theory of personal constructs can be claimed to cover all psychological phenomena – cognitive, conative and affective. It also refers back to itself, not like behaviourism, for instance, which has difficulty in accounting for the behaviour of the behaviourist.

2.6 Grid technique and the Rep Test

The general directives at the end of his Chapter 4 (Kelly, 1955) give a better idea of what Kelly was trying to achieve with grid technique than the detailed recommendations in the following chapters. It is introduced as a flexible method of interviewing for a clinician to adopt in studying his client's psychological problems. No prepared materials are needed: the grid can be devised by the clinician and the client working together and filled in as it is constructed. The terms should be adapted to the client's situation: the elements should be ones with which he is personally involved and the constructs ones he habitually uses in thinking about them. He is the measurer, evaluating his own elements in terms of his own constructs, and not something measured with the clinician's yardstick, an element assessed by comparison with others in the clinician's grid. The results should reveal the client's options, the directions in which he recognizes that changes in his situation can be effected.

To describe his procedure in detail Kelly presents it later in the form of the role construct repertory test, or Rep Test, and imposes many unnecessary limitations in doing so. The reader who is thinking of using the technique should notice that he refers to the test as an illustration and points out that it can be modified in various ways. If the user takes it as an exact model and copies it invariably in every detail, he will make grid technique a much less sensitive and flexible instrument of observation than it need be. Rather, he should adapt it to the needs of the particular case in his own practice, and regard the Rep Test as a device for introducing the technique to literally minded first-year college students – an experimental application of the theory of personal constructs they may be tempted to try out on one another in the lab.

The interviewer needs a deck of 3 by 5 inch cards, with a role title on each and a blank space beneath it, and a form for recording the informant's responses. The twenty-four roles on the cards include a teacher you liked, a teacher you disliked, your wife or present girlfriend (alternatively your husband or present boyfriend), an employer you liked working for, one you disliked working for, your mother, your father, etc. The cards are handed to the informant one at a time and he is asked to enter the name of someone who fits the role in the blank space on each, without using the same name twice.

After they have all been filled in the interviewer picks out three passes them to the informant and says, 'I would like you to tell me something about these three people. In what important way are two of them alike but different from the third?' He records the response on his form as a construct and notes the two elements to which it applies. Next he points to the odd card and asks, 'How is this person different?' The answer is noted as the contrast to the construct. Then the informant is asked to go through the rest of the cards and sort them out, according to whether the construct or its contrast fits them best.

The form for the responses includes a table with columns for the elements and rows for the constructs. The roles are listed above the columns to identify them and space is provided beside the rows for entering each construct with its contrast as it is elicited. The results of sorting the elements according to it can be recorded by putting ticks along the row for it under the headings for the elements to which it applies and leaving the cells for the contrasted elements blank.

The constructs with their contrasts are elicited one after another by presenting a different set of three cards each time, and the table, i.e. the grid, is filled in row by row. The number of constructs to be elicited is not fixed, but no informant's supply is inexhaustible; his limit is usually reached fairly soon.

2.7 Merits and limitations of the Rep Test

A complete construct system cannot be covered by one grid, for some parts are not commensurate with others. A subsystem of comparable elements had to be chosen as an example for the Rep Test. Kelly selected one of major interest to everybody. The topic of 'people I have known' might be described as a public highway into every private universe; as Pope said, 'The proper study of mankind is man', and the bipolar contrast to his aphorism is just as pertinent, as others have pointed out. But of course there is no implied guarantee that this route will lead to the region of greatest clinical interest in a particular case.

Entry into an informant's microcosm is effected when he picks the cast from his own acquaintances to play the roles for his version of the Rep Test. The performer he chooses for each role is someone who occupies a definite place in his construct system, and when he goes on to complete his grid he is no longer concerned with the roles — he refers to the performers and gives their locations. For instance, the position he assigns to his mother in his construct system indicates his personal feelings about her, which may not necessarily be at all like those sons are generally supposed to have for their mothers.

The procedure is intended for exploration, not for testing hypotheses. Instead of excluding variation as far as possible from all sources except postulated ones, it extends the scope for the appearance of unexpected phenomena in all directions. Once the roles have been specified, choices are left open to the informant as far as possible. The roles are deliberately diversified: they include men and women, contemporaries and elders, attractive and repulsive characters. Shifting the focus of interest by presenting a different set of three elements every time a construct is to be elicited offers continual opportunities for new ones to be expressed.

It is evidently a sound principle to diversify both the elements and the constructs as far as possible within the boundaries of the region to be explored. Presenting a different set of three elements every time a construct is to be elicited (the 'triad' method) is a good way of obtaining a series of constructs which do not overlap excessively; and it can be used equally well whatever the elements. Choosing acquaintances to fill the roles specified in the Rep Test provides diversity when the region to be explored is the informant's system for construing people but cannot be extended to other systems. Although these are methods worth recommending in some circumstances, other ways of obtaining the elements and constructs for grids evidently need to be considered.

In retrospect the omission of any form of self-concept from the elements in the Rep Test seems ill advised. A great deal of evidence has accumulated since, illustrating the value of examining relationships between 'myself' and other people included as elements in grids, and also between different self-concepts, such as 'myself as I am' and 'myself as I would like to be'. Much is to be found in the companion volume, *Explorations*, in the chapters describing clinical applications of grid technique.

The most peculiar limitation of the Rep Test is in the range of the scale provided for recording the variation between the elements in terms of a given construct. Kelly commits himself to the rule that it must always be a simple dichotomy, even to the extent of formulating it as the 'dichotomy corollary',

> a person's construct system is composed of a finite number of dichotomous constructs

and appears to expect it to be applied not in the Rep Test alone but in every instance of grid technique. It is an important rule to break. Constructs which only offer a dichotomous contrast are best avoided.

Kelly sometimes takes after Humpty Dumpty — his words mean just what he chooses them to mean. 'Corollary' certainly does not carry its generally accepted meaning in his text. The dichotomy corollary is not a logically necessary consequence of his fundamental postulate; it is simply a proposition he has tagged on because he considers it an important part of his theory. But though the theory is excellent in parts, like the curate's egg, this is not one of them. His reasons for including it are not quite explicit or cogent. Perhaps he thought it could solve the problems of psychological scaling discussed by Gutmann and others (Torgerson, 1958) or perhaps he was influenced by the use of binary notation in digital computers. These can only be conjectural explanations.

A pair of alternatives, yea/nay, present/absent, animate/inanimate, etc., makes the crudest of all distinctions. Twenty questions offering such alternatives are allowed to one side in the well-known parlour game to help them guess what the other side is thinking of, and often twenty is not enough. A common mistake is to ask virtually the same question twice over instead of introducing a fresh distinction each time. Similarly, more constructs will be needed to locate the elements accurately in a system where they are all dichotomous than in one where they provide finer distinctions, and more will be needed when they tend to duplicate one another, i.e. are correlated, than not. In this respect a Kelly-type construct system must be approximately the most inefficient one that will work at all, and it may have a special interest for that reason.

The system which is theoretically the most efficient is one using a set of continuous, independent variables, but it is virtually impossible for anyone to operate without electronic aids. The grid recorded by an informant is replaced by the most efficient one with the same contents when analysed by the method given in Chapter 8. The results provide the interviewer with the means to examine the contents thoroughly from end to end.

The contents are not modified in any way during the analysis, e.g. by adjusting them to fit some prior assumption about what they ought to show. If they were trivial originally, the results from the analysis will be trivial too. the risk can only be prevented by the interviewer's insight, patience and experience at the time when the grid is being constructed and filled in.

2.8 Data from personal questionnaires

Any form of questionnaire or inventory which can be filled in by an informant to record his mental state will generate a grid, provided it incorporates several scales and is administered on a series of occasions. The scales serve as the constructs and the states on different occasions as the elements. Naturally the scales should be sensitive to states which are liable to fluctuate and the occasions should cover a period when fluctuations are expected to occur. For instance, a grid might be built up from a patient's records of his mental state during a course of treatment for relieving acute anxiety or depression. The questionnaire might include experimental scales for measuring changes the treatment is intended to induce and control scales for measuring changes only expected to occur incidentally from external causes. Scales for assessing stable personality traits would presumably be irrelevant.

The technique Sharpiro (1961) has proposed for constructing personal question-naires is particularly appropriate for this purpose. A questionnaire is specially designed for each patient, with scales referring to the symptoms that, according to him, distinguish between his disturbed state and his normal state. A detailed account of the technique and ways of extending it appears in Chapter 14 of this volume and an analysis of some results in Chapter 7 of the companion Volume *Explorations*.

3

GRID TECHNIQUE IN PRACTICE

3.1 Preparing to construct a grid

It would be difficult to lay down any definite rules for constructing a grid without circumscribing the procedure unnecessarily. It is more practical to start from a central point and survey the possibilities without expecting to arrive at ultimate boundaries.

Suppose the grid is to be used simply for one informant on a single occasion. The setting is most likely to be a clinic, the interviewer a consultant and the informant a patient. Dresser (1969) describes a suitable procedure for a psychiatrist. Before interviewing the patient he studied the accumulated case notes. During the interview he went over them again, asking supplementary questions on every subject: the personal and the family history, the occupational history, the history of the present illness and previous ones, etc. So finally he was well able to judge who were the important people in the patient's life and what were the expressions he habitually used in referring to them, and found no trouble in producing a tailor-made grid, with a diversified assortment of people from the patient's acquaintanceship (including myself 'as I am' and 'as I would like to be') as the elements and his habitual expressions for them as the constructs. In an informal way he had succeeded in eliciting both sets of functions for the grid.

Even when the consultant has used such indirect ways to reach opinions about which elements and constructs should go into the grid, he may introduce it simply by proposing to talk about elements of one kind, leaving the patient the opportunity to choose particular examples of that kind. Rowe points this out in a personal communication, explaining that the order in which the patient gives them and his omissions and inclusions may reveal much about him.

> The psychologist may ask, 'Who are the people who are or have been important to you?', and find that the names he is given are not ones he would have expected from his prior knowledge of the patient. A woman might begin with her children and leave her husband until last. A man might not mention his family but list his boss and his mates. The omission or inclusion of parents tells something of the relationship existing between the patient and them. The psychiatrist is not usually included as an important person, even though the patient's life may have been affected enormously by the change from being a private person to becoming a patient. Questioning often reveals that the psychiatrist is seen as *Deus ex machina* without human characteristics. Sometimes the patient includes characters whose importance no psychologist

would have predicted. 'One patient whose case notes showed him to be a graduate of Oxford and a Battle of Britain pilot said 'The people most important to me are Beethoven, Shakespeare, Einstein, Wagner, Nietzsche, Gandhi, Schweitzer, Kant, Buddha, Rembrandt and Confucius . . . ' and so on for twenty famous names. He rejected the suggestion that his wife and child should be included, and the completed grid revealed a construct system of unsurpassed nobility. Subsequently investigations revealed that much of what he had claimed, including his degree and war record, were entirely imaginary.

The local interest of a grid may not be lessened if its elements are fictitious or deliberately misconstrued.

Similarly, the consultant may go through the procedure of eliciting constructs even though he has already formed opinions about the ones he would like to see included, asking what is special about one of the elements, or what is the main difference between two of them, or how two are alike and different from a third; and only supply constructs he thinks important at the last if the patient has failed to mention them spontaneously.

Dresser administered his grids at a second interview. The elements were listed down the left-hand side of a piece of ruled paper. A matching strip of paper, with a construct and its contrast as a heading, was laid alongside on the right and the evaluations of the elements were entered opposite them on the strip. They were recorded on a seven-point scale. When the strip had been filled in it was removed and another headed by a different construct was substituted. The patient was left with no opportunity to compare the entries he had put on different strips. Dresser used about twenty-five elements and constructs in most of his grids and they took up to about 45 minutes to fill in.

The task of constructing a grid should not be approached prematurely. To be useful in a clinical setting it needs to be focused on the psychological problems presented by the patient, and the more information the clinician has access to and the closer his rapport when he proposes making a grid the more likely is it to provide valuable evidence for diagnosis and therapy. He should start with some general idea of the area he intends to study in his patient's inner world, express his interest in it and obtain indications of the elements and constructs which cover it by open-ended questions. When communication is well established, he is familiar with his patient's language and has already identified some of the most suitable terms for use, he may propose putting them together in the form of a grid and invite his patient's help in constructing and applying it.

While these considerations are all in favour of collecting plenty of background information before going on to construct and administer a grid, there are other considerations in favour of avoiding delays in introducing the subject. Rowe recommends completing a grid at the end of the first session with an out-patient, adding that they will discuss the results together at the next. This attracts the patient to keep the next appointment. More than one grid may be used during a course of treatment, so omitting important material from the first may not matter much.

Making a grid is an experiment in which clinician and patient collaborate. The results can often influence the progress of the case. Patients have reported gaining insight from doing it and clinicians have used the results for discussions in psychotherapeutic sessions.

But opinions also differ about the advisability of showing the patient the results from analysing his grids. Feldman (1972), for example, decided not to do so when using the same grid repeatedly during a course of treatment, in order to avoid having the changes recorded on succesive occasions influenced by the patient deliberately. This dedication to scientific impartiality unfortunately made it more difficult for him to secure the patient's continued collaboration. The priorities of research and treatment can often conflict.

3.2 Numbers of constructs and elements

Dresser's grids were exceptionally large. In a trial series of just over one thousand grids from miscellaneous sources the modal number of constructs was found to be fifteen; only 1 per cent contained fewer than six and only 5 per cent over twenty-five. For elements the range was rather narrower, with 0.1 per cent. under six, and only 2 per cent over twenty-five; the modal number was twelve.

A grid with n constructs and m elements will contain nm entries altogether; there will be $n(n - 1)/2$ correlations between the constructs and $m(m - 1)/2$ measurements of distance between the elements; the component-space will be limited to n or $m - 1$ dimensions, whichever is the smaller.

A small grid, say one with no more than fifty entries, may yield a few measures describing its properties as a whole, serving to compare it with other grids in a series; that is to say, it may do for some nomothetic purposes. But internal disparities may occur even in such small grids, making global measurements misleading. Repeating one to measure its consistency may show that almost all the inconsistencies in it are due to the informant's indecision about how one of the constructs should be applied or one of the elements evaluated, and not to any diffuse uncertainty or disorder.

If a grid is to sustain idiographic conclusions independently of other sources of information it should range over a wide enough region and contain enough data to permit comparisons between different parts of it. A 10 by 10 grid will provide forty-five measurements of correlations between the constructs and the same number of distances between the elements. This begins to be enough to examine which are the closest. In a child's grid referring to people, for instance, we may see whether he identifies himself more closely with his mother or his father, or measure how far, in what directions, 'myself as I am' deviates from 'as I would like to be'.

It is an advantage to have an approximately square grid — ideally one with $n = m - 1$. If the component space is limited by its construction ot less than n dimensions the correlations cannot all be independent of one another and, similarly, the distances between the elements cannot all be independent if it is restricted to less than $m - 1$ dimensions.

3.3 Choosing elements

In keeping to people from the patient's acquaintanceship for his elements Dresser may have followed the model of Kelly's Rep Test unnecessarily closely. Such elements are appropriate, of course, when the patient's psychopathological condition is associated with difficult interpersonal relationships. But statements simply attributing a quality to a person, such as 'A is generous' and 'B is mean', may be rather inadequate for expressing them. The use of dyadic relationships as elements (Ryle and Lungi, 1970), which is exemplified in Chapter 4 of *Explorations*, provides greater scope. A's relationship to B may differ from his relationship to other people, and from B's relationship to A, etc. It may be intended in one way by A and interpreted in different ways by others. It makes an element of a kind an informant may enjoy assessing in a grid.

Objects can fill the role of elements just as easily as people. Table 3.1 gives a grid obtained by Mitcheson (personal communication, 1968) from a drug addict. The

TABLE 3.1 An example of a grid showing an addict's evaluations of a range of drugs (the elements) in terms of their effects (the constructs)

| Constructs | Elements | | | | | | | | |
	A	B	C	D	E	F	G	H	I
1	2	2	4	1	1	4	5	2	1
2	3	1	2	1	1	2	1	1	2
3	3	1	2	1	1	2	1	1	2
4	3	1	2	5	5	2	5	1	5
5	2	1	2	1	4	2	3	1	3
6	2	1	2	5	5	2	5	1	4
7	3	3	1	3	2	2	1	3	2
8	3	3	2	3	3	3	3	3	3
9	3	1	4	3	2	3	4	2	2
10	3	2	2	2	1	2	1	2	4
11	3	1	1	1	3	1	1	1	3
12	3	5	2	1	1	5	3	5	1
13	2	5	2	4	3	5	3	5	4
14	3	3	3	2	1	3	2	3	2

Key
Elements
A. Alcohol
B. Barbiturates
C. Cannabis
D. Cocaine
E. Drynomil
F. Heroin
G. L.S.D.
H. Mandrax
I. Methedrine (injections)

Grades
Causes this effect
1. very strongly
2. to some extent
3. neutral
Causes the opposite effect
4. to some extent
5. very strongly

Constructs
1. Makes me talk more
2. Makes me feel high
3. Makes me feel blocked
4. Makes me feel sleepy
5. Gives me a warm feeling inside
6. Makes me feel drunk
7. Makes me imagine things
8. Makes me feel sick
9. Makes me do things without knowing what I'm doing
10. Helps me enjoy things
11. Gives me a good buzz
12. Makes me tense
13. Makes me feel sexy
14. After taking it I may see or hear people who aren't really there

elements are drugs he has tried and the constructs effects they produce. It may serve as a rather freakish example of the kind of data often collected when grid technique is used in market research.

In other cases other choices of elements may be more suitable. In obsessions and phobias, for instance, the patient's problems are often concerned more with situations than people. A set with varied affective tones may be elicited by unstructured exploratory interviewing. For example, Watson took situations as the elements in the grid obtained from a self-mutilant girl, which serves as the example in Chapter 9; the constructs are possible consequences. Admittedly the elements might also be described as self-images: when asked to evaluate a situation such as wanting to talk to someone and being unable to, the girl must imagine herself, not anyone else, in that situation and indicate the effect it would be most likely to have on her personally. But 'myself' could be detached from the situations if desired by asking the informant to complete the grid twice, once for herself and once as 'most people would'.

More surprisingly, situations are used as the constructs in the grid used by Watson, Gunn and Gristwood for their study of prisoners serving long-term sentences in *Explorations* (Chapter 13). They were 'situations which a recidivist prisoner might meet in his stress-filled life, portrayed to each subject by narrative as vividly as possible'. Nine ways of responding, such as getting drunk, feeling tense and punching out, served as the elements. Each was printed on a separate card. The prisoner ranked them be picking out the one he would be most likely to adopt in the situation, then the most likely one from the remainder, and so on. After this another situation was described and the responses were ranked again.

3.4 Transposing elements and constructs experimentally

Since situations were used as elements in one of these grids and as constructs in another, the elements and the constructs in a grid must be interchangeable — sometimes at least. Or more precisely, the same set of terms may function as the operators in one grid and as the operands in another; one cannot expect that interchanging them will leave the contents of the grid unaffected.

The experimental procedure will need to be modified. In the experiment with the prisoners, for example, all the situations would have had to be described before a prisoner could be asked to rank them in terms of the first construct. Their vividness and dramatic details would have begun to fade before he had started filling in the grid, and grown more dim and vague, unequally, as he went on. Obviously the procedure actually adopted was more suitable for the experiment.

However, the information obtained will be different, as may be seen by comparing the girl's grid with the prisoners'. In both of them the elements are ranked in terms of the constructs. Consequently the entries for every construct in the grid have the same mean. In the girl's grid, where each construct refers to one of the possible consequences of being in the given situations, there is no evidence to show whether some consequences are more likely to occur than others. But in the prisoners' grid, where the situations are constructs and the consequences of being in them are the elements, evidence of this kind is obtainable. Watson, Gunn and Gristwood have used it effectively in *Explorations* (Chapter 13). They succeeded in showing that recidivists with records of gambling and alcoholism could be identified by their predilection for certain responses, particularly in certain situations.

Terms can also often be transposed singly. For example, 'myself' may be included as one of the elements in a grid along with other people, or it may appear in the form 'like me' among the constructs, leaving the other people still as elements. Here again transposing must change the structure of the grid and produce different results. Which is the more suitable way of presenting the terms will depend on what results are of greatest interest.

3.5 Connections between personal construct theory, logic and probability theory

Enough instances may have been introduced by now to sustain the conclusion that the typical proposition in personal construct theory, *E may be construed as C*, paraphrases the typical proposition of Aristotelian logic, *S is P*, i.e. subject is predicate. It is difficult to conceive of any proposition that can be stated in one of these forms and not the other.

Transposition operates similarly: *C applies to E if E is construed as C* just as *P applies to S if S is P*. Both statements differ in the same way from the algebraic equation $x = y$; for this implies that the universe of all possible values of x coincides point-to-point with the universe of all possible values of y, whereas 'S is P' only implies that S is a member of the set to which P applies and likewise 'E can be construed as C' only implies that E falls within the range of convenience of C.

The proper distinction between the two is that 'S is P' is an objective statement while 'E can be construed as C' is subjective. Valid reasoning or sufficient evidence may prove in particular instances that the proposition 'S is P' is true or false beyond all possible doubt whatever. If it is not meaningless, its truth value can only be 1 or 0 in every instance, although which may be uncertain for lack of conclusive proof. On the other hand, 'E can be construed as C' only connects E and C within some real or imaginary personal construct system.

Every entry in a grid is the numerical formulation of a proposition which has some psychological meaning for the informant. It cannot be rejected as a datum on the grounds that it is untrue, or that no one else agrees with him, or that he should never have put it on record or never intended to do so, or has never expressed the same opinion on any other occasion. No jot or tittle of a grid can be dismissed as devoid of meaning, though much of it may be redundant. And the grid as a whole is much more than a simple list of statements; it reveals a complete nexus or interrelations between them.

Probability theory, too, has a subjective basis, as Keynes (1921) originally pointed out. It presupposes an observer with insufficient evidence endeavouring to reach an opinion whether 'S is P' is true or false, and offers estimates derived from assumptions as substitutes for facts. Its discourse employs subjective terms systematically. It discusses such questions as how to define the best estimate and determining its confidence limits. It provides a continuous scale for measuring probability with the range from 1 for positive certainty through 'fifty, fifty' (i.e. 0.5) for complete uncertaintly to 0 for negative certainty. Take *'Oxford will win the next boat-race'*, for example. At any specific time before the event someone may argue from all the available evidence that the most reasonable estimate of Oxford's chance is some specific quantity between 1 and 0. He assigns the element a location on the statistician's scale. In this case the evidence is not easy to quantify and the conclusion is obviously an expression of opinion. In other cases there may be practically no doubt about the best estimate. Thus the chance of turning up the ace of spades by cutting a deck of cards may be but at 0.02 (correct to two decimal places), provided the dealer and the deck are above suspicion, and it will be just the same wherever the deck is cut. Reasonable though this may seem it is not a statement of fact but a highly sophisticated expression of complete uncertainty. It cannot actually be true.

3.6 Methods of presentation and scaling

It is possible to choose the constructs for a grid first and elicit the elements afterwards. For example, data in the form of grids were obtained in an experiment of unrecorded origin by providing five constructs:

1. I do/do not
2. I can/cannot
3. I want to/do not want to
4. I ought to/ought not to
5. I have to/do not have to

and asking informants to specify activities involving some conflict between them. Elements elicited included such activities as *complete my income-tax returns, have a steady job, keep my room tidy, visit my parents, spend money on beer, make a living out of writing, have a feeling of release, sail a boat, ask myself what my life is for*, etc. A five-point grading scale was used to indicate how they were evaluated in terms of the constructs.

So this experiment began with a given set of constructs and a fixed grading scale, and as it went on the list of elements grew longer. The effect is the same with the semantic differential. Although the elements (concepts) are supplied by the experimenter and not elicited from the informant, they are presented to him one at a time for assessment on a prescribed set of constructs (scales) with a fixed metric, so the experiment proceeds by extending the list of elements.

With grid technique the experiment usually extends in the other direction. The elements and the method of scaling are given at the start and the constructs are elicited one after another, forming a lengthening list as the experiment continues. The alternatives may be described as presenting one element at a time or one construct at a time.

Whichever way the experiment extends the method of scaling should be decided before it begins, and it is for the experimenter to take the decision. He may adopt one of many well-tried methods or devise one of his own. If he intends to use grid technique regularly with a certain class or informant, he should try several methods to find one which will satisfy both him and them; it should provide him with information of the kind he needs and they should be able to use it confidently. Too often decisions are taken arbitrarily without any trial runs to judge the merits of different possibilities. The experimenter interviews patients of all kinds in a clinical setting so he may vary his methods according to his opinion of their potentialities. What suits one may not suit another.

Changing methods during the course of a grid is inadvisable. Informants take the idea of using the same scale for all their constructs so readily that any alteration interrupts communication. It would be possible, however, if rapport is good, to ask an informant after he has completed his grid to do it again using a different method of scaling and then compare their convenience.

Virtually any number of points may be provided on grading scales, from two up. Kelly (1955) advocated simple dichotomy; many of his followers keep to it out of respect for personal construct theory, where the proposition that all constructs are dichotomous is enshrined as a corollary. An experimenter might extend his scales without forsaking the theory by asking the informant to dichotomize each construct twice. For instance, if he says that something is sweet, in contrast with something else which is sour, he could be asked, 'Very sweet, or just rather sweet?'. No example of this method of making a four-pointed scale has been noted, but Shapiro has developed a more complicated method based on paired comparisons for his personal questionnaire (see Chapter 14 and *Explorations*, Chapter 7). Topçu (1976) found no difficulty in getting his informants to use a six-point scale beginning with a dichotomy and following it up with a choice of three points: the supplementary question for 'sweet', for instance, would be 'Very sweet, quite sweet, or only slightly sweet?'.

Most experimenters, however, are in favour of providing a neutral grade at the centre of their scales, with an equal number of grades above and below it. Any such scale will have an odd number of grades altogether. The most restricted is a three-point scale, allowing a place between sweet and sour, for instance, to locate drinks which are merely dry. Five-point scales, which come next, are the ones most generally favoured in opinion surveys and market research. The amount of variation they record is enough for most experimental purposes and informants usually find them easy to operate. Even flying instructors in the R.A.F. during the war, who made a cult of laconic understatement, could be persuaded to use a five-point scale for grading their cadets. It went: above average, good average, average, poor average, below average. A more generally suitable gradation for a five-point scale, say from good to bad, would be: very good, rather good, not particularly good or bad, rather bad, very bad. Another would be: good, more good than bad, neither good nor bad, more bad than good, bad.

A seven-point scale such as is used in the Semantic Differential is widely preferred for grids, but is not much more sensitive to variation in practice than the five-point scale. Informants tend to ignore grades 2 and 6 and thus reduce it to five points. Osgood defined his seven grades as: extremely X, quite X, slightly X, neither X not Y or equally X and Y, slightly Y, quite Y, extremely Y, X and Y being a pair of contrasted terms.

Stephenson's elaborate methods for operating scales with a wider range of grades (mostly eleven, but occasionally nine or thirteen) and prescribed distributions have not found much favour among later users of grid technique. Presumably they make excessive demands on the patience of the informant and the ingenuity of the interviewer. Stephenson's informant took fifteen sessions to complete the grid with sixty elements mentioned above, making a Q-sort of them in terms of one construct on each occasion.

There is really no need to take as much trouble as he did to ensure that distributions fit a desired form — for two reasons. The first, which is sufficient in itself, is that the method given here for analysing grids does not involve any assumptions about what forms their distributions take; it applies equally well to all. It calculates successive estimates which approximate closer and closer to the observations until an exact fit is obtained. The criterion for calculating them is that the sum of the squares of the differences between them and the observation must be minimized at each stage. Secondly, even if it were a necessary condition for the analysis that the residual deviations should be normally distributed it would be unnecessary to coax the original data into that form. The tendency for such deviations to converge on the normal form, which is proved by the central limit theorem and confirmed by observations from innumerable sources, is as conspicuous in grids as anywhere else. Whatever the form of the original distribution, once the variation from the main source is extracted, the distribution of the residual variation approximates to the normal form. This can be verified with the INGRID program, which has a special option for printing out a detailed list of deviations at various stages during the course of an analysis.

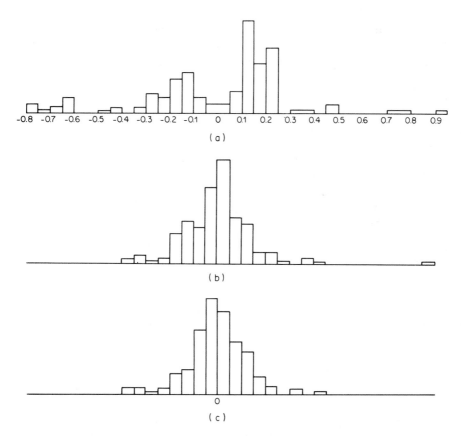

FIGURE 3.1   Distributions formed at various stages in the analysis of a grid (using data from *Explorations*, Chapter 8). (a) Normalized variation about the construct means. (b) Residual variation after the extraction of one component. (c) Residual variation after the extraction of two components

The grid recorded by Rowe as T1 in *Explorations* (Chapter 8) could hardly be bettered as an example of non-conformity to all known forms of frequency distribution. Figure 3.1 shows how its residuals approximate to the normal form as the analysis proceeds.

The statisticians' infinitely divisible scales expressed as percentages or proportions are of course ideal theoretically and have often been tried for recording grids. But they seldom work well in practice. Too often what happens is that one or two of the elements get thrown out with grades of 0 or 50 and fine distinctions are drawn between the rest over the range between 90 and 100. If such a distribution recorded the informant's assessments of the elements accurately, there could be no objection to accepting it, but the impression it generally gives is that he cannot manage the measuring instrument properly and should have been given an easier one to handle.

3.7 Normalization

Every scale used in a grid is limited at its top and bottom and has a finite number of intervals, which are only roughly equal, in between. It is like an ill-made flight of stairs, with unequal treads; all one knows for certain is that each step up is nearer the top and further from the bottom. Measurements on it can be recorded in integers, i.e. by counting the number of steps. This is how ranking works, too.

Though the interviewer may have reasons for supposing that the measurements he is recording are inaccurate, he cannot generally expect to improve on them by altering them. If he did he might be accused of tampering with the data; on the whole it is better to accept them at their face value.

One method that may be considered in certain circumstances is normalization. It is used in statistics as part of the procedure for calculating correlations between variables. Their correlation is obtained from their covariance by scaling their variances down to 1 and their covariance proportionately. It is a more convenient statistic to consider than the covariance when the variables are not commensurate, and is habitually applied to psychological tests because of their arbitrary incommensurate variances.

The experimenter may consider normalizing the constructs in a grid if he doubts whether they are really commensurate. The drug addict's grid (Table 3.1) is a good example. The instructions are to evaluate the effects of the drugs on a scale which runs from 'causes it very strongly' to 'causes the opposite effect very strongly'. It may be reasonable to suppose in general that while some drugs produce certain effects others may counteract them, but the patient has not reported that any of the drugs produce effects which are the opposite of

2. Makes me feel high
3. Makes me feel blocked
7. Makes me imagine things
11. Gives me a good buzz
12. After taking it I may see or hear people who aren't really there

These constructs only record variation in one direction; the rest record variation in both directions except 8, 'Makes me feel sick', which hardly records any variation at all.

Should the constructs be normalized or not? The question is easier asked than answered. The grid could be analysed in both ways, using the option to normalize which is included in the INGRID 72 program, to see what the differences are.

Normalization gives an equal weight to the variation recorded by every construct in a grid and may be considered desirable for this reason. Or rather, there may be reasons for considering that some constructs in a grid should not be given more weight than others. Thus the question whether to normalize or not only arises when there are wide differences in the total variation of the constructs, and then the decision should rest with the person responsible for conducting the experiment and interpreting the results. If he considers the differences psychologically important, he may decide to retain them throughout the analysis of his grid and therefore prefer not to have it normalized. Or yet he may be satisfied with the information he obtains about the differences in means and variation of the constructs in the preliminary stages of the analysis and still assign them equal weights when their covariation is analysed.

Constructs with the largest variation may not necessarily be those which show the clearest discrimination between the elements. If the informant only uses the two extremes of a seven-point scale for one of his constructs, and the whole range for another, the former may well have more variation though it is virtually only a dichotomy. The time to prevent this occurring is when the grid is being recorded. If the interviewer is using a seven-point scale he should try to avoid including a construct which is merely dichotomous by asking supplementary questions in order to arrive at another parallel construct in terms of which finer distinctions can be made (see Section 3.10). Otherwise the best that can be done is to take the option to normalize and thus prevent the cruder construct from having the preponderant weight. If it is quite clear that some of the constructs are incommensurate with others, normalization is unquestionably justified.

3.8 Ranking

Ranking is another method of scaling which has frequently been used and found satisfactory. The universal practice is to show the informant all the elements and ask him to rank them in terms of one construct at a time. Presenting the elements one at a time is practically inconceivable.

There is evidence to show that ranking is as reliable as more elaborate methods of comparison and tends to maintain a constant level of reliability, dependent on the informant, for different numbers of elements over the whole range used in grids. Informants are glad of the opportunity to survey all the elements before beginning to rank them, find the task simple, do it quickly and feel satisfied with the results (Slater, 1960, 1965).

Salmon (*Explorations*, Chapter 2) has found the method suitable for young children. Gunn and others used it successfully with prisoners serving long-term sentences and Spindler Barton recommends it for low-grade defectives (see *Explorations*, Chapter 2, 3 and 13). Each simplified the task for their informants by instructing them to pick one at a time. All the elements were presented when a new construct was introduced and the informant was just asked to pick out the highest on the scale; then it was put aside and he was asked again to pick out the highest from the remainder, and so on until only the lowest was left. Abler subjects whose interest is better sustained may be shown the set of elements and simply asked to order it according to the construct in their own way.

The convenience of ranking in the interview is offset by two constraints it imposes on the data, which reduce the amount of information that can be extracted from the results. One is that it obliges the informant to distinguish every element from all the others in terms of every construct. If he does not the grid must be treated as one of gradings and not rankings. The other is that all the constructs in the grid must have the same mean and variance, since one ranking of m elements can only differ from another in being a different permutation of the first m natural numbers. The mean is always $(m + 1)/2$ and the total variation about it $(m^3 - m)/12$.

Forcing the informant to distinguish between all the elements in terms of every construct may exaggerate differences between them and suppress similarities that might be of psychological interest. For instance, in reporting on her patient's second grid, recorded after his treatment had been completed, Rowe and Slater note (*Explorations*, Chapter 8) that he does not distinguish at all 'between his feelings for his psychiatrist and the girl he is going to marry. They are both perfect in every respect'. The girl's opinions about this and the prognosis for the marriage are outside the scope of the study. The point to notice here is that the observation could not have been made if the elements had been ranked instead of graded.

Evidence is also obtainable from comparisons between the means and variances of constructs when they are free to differ as they are with grading. For example, in comparing the patient's grids before and after treatment Rowe and Slater comment, 'The changes in the mean and variance of construct 8 show that his guilt feelings are greatly diminished all around and no longer play an important part in determining his attitudes to other people or his estimates of their attitudes to him.' Another striking example is to be found in a paper by Orley and Leff (1972), reporting results of a research in Uganda. They used illnesses as elements and traditional ways of distinguishing between them as some of the constructs in a grid completed by trained paychiatric nurses, student nurses and secondary school pupils of about the same educational standard, to study the effects of training. They found among other things that whereas the school leavers drew wide distinctions between illnesses that come of themselves and illnesses that are sent by sorcery the student nurses made far less distinction and the trained nurses very little indeed. The variances in the three groups were 33.4, 24.0, and 4.2. Similarly results were found with the construct Ganda/European.

If an experimenter is wondering whether to adopt ranking or grading as his method of scaling he should consider whether he needs to obtain evidence of such kinds for his research as well as what degree of cooperation he can expect to gain from his informants.

3.9 Geometrical scales

The idea that geometrical methods may be applied to subjective measurements is the starting point of this book. There is no need to discuss whether it is true or false; the critical question is whether the methods proposed are useful — and that may be left for the reader to decide after finishing the companion volume, *Explorations*.

The geometrical notion of a scale is an ideal to which subjective scales may approximate, more or less. It is simply a straight line. Two points, say A+ and A−, are enough to define it. Though they must be separate they need not be far apart. The line between them, i.e. the axis A±, may be said to run either from A+ to A− or, in the opposite direction, from A− to A+. It may be extended indefinitely in either direction and it is infinitesimally divisible.

To find the distance along the axis of two points, X and Y, which are not actually on it, perpendiculars may be dropped from then onto it, and the length of the intercept may then be measured. The length will be the same on any other axis, say Q±, which is parallel to A±. So if it is inconvenient to measure the distance between X and Y on A± another parallel axis Q± will do instead; and for Q± we may choose an axis which passes through a third point, O. Thus to measure the distance between two points in any number of different directions we may choose axes which all pass through a common origin or centre at O, radiating out from it in different directions. If a line is known to pass through O, one other point is sufficient to define its direction.

Q+ and Q− are bipolar opposites if Q+ is opposite Q− on the other side of O. And as every axis through O can be extended indefinitely it must be bipolar. To conclude, distances in any multivariate dispersion of points can be measured by a set of bipolar reversible axes with a common origin.

3.10 Choosing constructs

A distinction is often made between supplied and elicited constructs, the former being ones the interviewer provides without any suggestion from the informant and the latter being ones the informant provides without any suggestion from the interviewer. In a clinical situation, if there is a continuous exchange of ideas when the grid is being drawn up, the distinction is apt to become blurred. It is in a large-scale experiment, when a number of informants will be asked to complete it, that a grid with supplied elements and constructs is most likely to be used. Even then some exchange of ideas may have occurred. Group discussions or loosely structured interviewing will probably have been conducted previously with a comparable sample of subjects in a pilot study to elicit suitable terms. If not, the results may be disappointing.

Completely spontaneous constructs are most likely to be elicited by Kelly's method of taking three elements at a time and asking the informant to consider them in isolation from the rest and described some way in which two are alike and different from the third, or Landfield's method 1971, (*Explorations*, and cf. Chapter 6) of presenting only two and simply asking the informant to describe some difference between them.

Such constructs may not always be suitable for inclusion in the grid in exactly the form in which they are originally given. The constructs to be included should

(a) show marked constrasts between some elements and others,
(b) apply to all the elements in some degree,
(c) diverge in content.

Without substituting his own constructs the interviewer may adapt unsuitable constructs to these requirements by asking the informant to amplify and explain them. Keeping the notion of a geometrical scale in mind he may regard his task as one of replacing the construct first proposed by another parallel to it, distorting its meaning as little as possible. The need to do so may be rather more urgent if the elements have been selected and the method of scaling has been decided before the constructs are elicited.

At a preliminary interview choosing the constructs may be combined with choosing the elements. If the informant keeps picking out elements of only one kind in the region of interest, he may be invited to mention some of the other kinds and then discuss differences between them. In talking about holiday resorts, for instance, he might be encouraged to mention ones he dislikes as well as ones he likes, and discuss other people's preferences as well as his own. If a grid is drawn up in this way there is no need to adopt any formal procedure for eliciting constructs.

Moreover, the interviewer need not pick out two or three of the elements and isolate them from the rest when asking the informant to compare them. He may point out one and ask which of the others differs from it most, and then go on to enquire what the difference is. Or if this turns out to be difficult he may ask which of the other elements is most like the one he has pointed out and then enquire into the similarity. This procedure is less likely to elicit a construct which fails to include all the elements within its range of convenience.

If the construct he elicits does not extend to all the elements or makes a distinction too crude to be calibrated on the scale in use for the grid, what course of action should he follow? Suppose an informant comparing people said, 'These two are school-teachers; he's a mechanic'. It would be baffling to try to graduate such a distinction or extend it to people playing other roles. Should the interviewer simply forget it and ask for another difference? He may suspect that the objective distinction masks another which is subjectively more important. If so, he may ask supplementary questions in an effort to uncover it. He might say, for instance, 'What's so special about school-teachers? Why is the mechanic different?' and get a reply such as 'School-teachers are just bossy; he'll try and help you if he can'. He may then surmise that bossy and helpful are not just bipolar opposites but two constructs that cut across each other. So he may ask the informant to rate the elements from 'not at all bossy' to 'extremely bossy' and again from 'not at all helpful' to 'extremely helpful', in the hope of eventually uncovering the subjective implications of the distinction between teachers and mechanics.

It should not be assumed, however, that objective, factual distinctions are out of place in grids. Some people prefer to stick to the facts. For a character sketch of an acquaintance they may produce a lavish collection of reminiscences revealing him in an admirable or despicable light and then confine themselves to a terse triviality in conclusion: 'That's the sort of man he is.' The item of information, that 'the only paper Jack ever buys in the Daily Express', is enough to make tentative estimates of his political opinions, his social affiliations and the level of his intelligence and his income. It might be incorporated in a grid as the construct 'Buys the Daily Express', applying to people as elements, and calibrated on a scale from *regularly* to *never*.

The informant who goes on evaluating the elements in much the same way with constructs that only differ nominally presents the interviewer with a different problem: to discover whether his construct system is as one-dimensional as it appears. If he himself repeats the same procedure and rewards the informant with renewed interest every time a similar evaluation of the elements recurs, he may reinforce the informant's behaviour pattern and prevent any different point of view from emerging. It may be advisable to let the conversation become side-tracked and roam around the region of interest in an undirected way, and then follow a different method for eliciting constructs when returning to the task. There are a great many options open to the interviewer, and he need not feel committed to one in an individual case study.

A construct is best defined by two terms to specify the opposite ends of the scale it forms, for if they are remote enough the bipolar contrast between them will indicate the axis of the scale comparatively accurately however vague the meaning of each on its own. Words in common use are seldom precise. They have roots and branches like banyan trees, amidst which the main stem gets lost. They may sprout new meanings in slang or professional jargon. Many acquire an ambience of poetical associations and some are obviously ambiguous. They are not very suitable for scientific purposes. Thus *light* may be contrasted with *dark* or *heavy, old* with *new* or *young, soft* with *loud* or *hard, fair* with *dark, unjust* or *stormy, practical* with *theoretical* or *useless*, and so on. The ambiguity is lessened when both the contrasting terms are given.

At worst, a single term like *bossy* or *helpful* is sufficient. It serves as one pole, namely the emergent pole of a construct which has its opposite, latent pole left undefined, just as a single point may be used to specify a geometrical axis if it indicates a direction away from a zero origin. The use of a single term is particularly easy with ranking. The interviewer could hope to get all his elements ranked for bossiness simply by asking 'Which is the bossiest of these?' in the pick-one-at-a-time way.

3.11 Connections between functions

When constructs are elicited one may lead to another, and in a flexible interview constructs may suggest elements as well, just as elements do constructs. Suppose an informant evaluating acquaintances mentions that some of them make him angry. After noting this the interviewer may ask, 'What sorts of people make you angry?', and thus elicit further constructs, such as 'People who are rude' or 'mean', 'conceited', 'obstinate' or 'cruel'. Next he may ask, 'Do you know anyone not on this list who is particularly rude, mean or conceited?' or, alternatively, 'polite, generous or modest?'. Thus the grid may be amplified by additional elements as well as constructs.

Supplementary questions may also be used to obtain evidence outside the scope of the grid. After the informant has evaluated his acquaintances on the scale for 'make me angry/don't', the interviewer may ask, 'Why does so-and-so make you angry?', referring to the acquaintance placed highest on the scale. The reply might be, 'He tells lies about me and gets me into trouble.' This could lead to questions such as, 'When did he do that?', 'Who did he tell?', 'What did he say?', 'What happened then?', extending the interviewer's knowledge of the informant without necessarily amplifying the grid.

'Probing' is the technical term most commonly used for obtaining additional information during an interview by asking supplementary questions. A particular kind of probing is known in the literature of personal construct theory as 'laddering'. The term was introduced by Hinkle (1965) in an investigation of the implications of constructs in accordance with Kelly's orgainzation corollory. The following account of his procedure is summarized from the report on his research by Bannister and Mair (1968):

> Each informant was asked to give the names of nine people who currently played an important role in his life and whom he knew well. These served as the elements in a grid with the addition of his own name as the tenth.
>
> Constructs were elicited by the triad method, self being one of the elements included in each triad; thus all the constructs were self-referent. Moreover the informant was asked which pole of each construct described the kind of person he would prefer to be; and the construct was only retained if he indicated a definite preference for one of its poles.
>
> After ten such constructs had been elicited each was taken as the starting-point for supplementary questions. The informant was asked, 'Why would you prefer to be at this pole than that?', 'What are the advantages of this side in contrast to the disadvantages of that?'. The answer provided another construct, also with a preferred pole, about which the same question could be asked again. Thus a chain of connected constructs was derived from each of the original ones.

Applying the terms from Kelly's organization corollary, Hinkle described the original constructs as subordinate and the ones derived from them by his procedure as superordinate. This choice of words imparts a sense of direction to the sequence of questions. 'Laddering up' seems an appropriate expression for deriving successively more superordinate constructs from ones subordinate to them. Conversely, 'laddering down' would be asking questions to derive subordinate constructs from superordinate ones. The question, 'What sorts of people make you angry?', would initiate a process of laddering down.

Kelly's organization corollary is:

> Each person characteristically evolves for his convenience in anticipating events, a construction system embracing ordinal relationships between constructs.

In his notes on it he explains

> Not only are constructs personal, but the hierarchical system into which they are arranged is personal too. It is this systematic arrangment which characterises the personality, even more than the differences between individual constructs. . . . One construct may subsume another as one of its elements. . . . When one subsumes another its ordinal relationship may be termed *superordinal* and the ordinal relationship of the other becomes *subordinal*.

He takes the construct intelligent/stupid as an example to illustrate his use of the word 'subsumes':

> The construct good/bad may subsume the two ends of the intelligent/stupid dimension. 'Good' would include all intelligent things plus some which fall outside the range of convenience of the intelligent/stupid construct. 'Bad' would include all the stupid plus some which neither intelligent nor stupid. Intelligent/

stupid could also be subsumed under the construct evaluative/descriptive. It would be identified as evaluative in contrast with other constructs such as light/dark, which might be considered descriptive only.

(These notes have been summarized, see Kelly, 1955).

The notes include some reservations, such as 'The ordinal relationship between constructs may reverse itself from time to time'. And it might be argued that other corollaries, particularly those on range experience and modulation, go far to diminish the effect of the organization corollary.

Even so, it seems to me to be stated overconfidently. It appears to amount to an assertion that taxonomic principles must apply, or at least that they ordinarily do apply to construct systems, the relationship between superordinate and subordinate constructs being parallel to the relationship between genera and species. Admittedly human beings may be capable of organizing their construct systems hierarchically and some, perhaps, may deliberately attempt to do so. But a system that has been thoroughly organized hierarchically would seem to me one that has become ossified and ill-adapted to the needs of everyday life and intelligent discussion.

For centuries eminent scientists have laboured to develop taxonomic systems, notably for botany and zoology, but none have proved universally acceptable, and during the last century evolutionary theories and other developments in biology have tended to blur the borderlines between genera and species. Problems of taxonomy are no longer regarded as urgent.

The proposition that superordinate constructs have more implications than subordinate ones, quoted by Bannister and Mair from Hinkle, is contrary to taxonomic principles. Species are distinguished from other species belonging to the same genera by the possession of peculiar attributes or combinations of attributes in addition to the attributes common to all the members of their genus. Consider for example the classical syllogism, 'All men are mortal, Socrates is a man, etc.' The class of men is subordinate to the class of mortals and Socrates is subordinate to the class of men. Consequently, being a man has far more implications than being mortal, while being Socrates has so many more implications than merely being a man that his memory has already been preserved for over two thousand years. Taxonomy recognises that the particular specimen has additional attributes which distinguish it from other members of its own species. Definitions of species and genera are derived by processes of abstraction.

If the construct rude/polite were taken as a starting point for laddering up and the informant had said he would personally prefer to be polite, he might go on to answer the question 'Why?' by saying that polite people are considerate. Then at the next question he might say that considerate people are good-natured. Returning to the original list of people included in the grid he might afterwards admit that some people who are polite may not be good-natured and some who are good-natured may not be polite, or, rather, that some people are more polite than good-natured while others are more good-natured than polite.

Grids do not generally provide any evidence of ordinal relationships between constructs. The best chance of obtaining any evidence of them is when there are only a few elements, little to choose between them and only a dichotomous scale available for evaluating them, and when the informant is dull-witted and the interviewer goes on pressing him for further constructs after his repertoire has become exhausted. When a wider variety of elements is introduced and a more sensitive scale is applied, any evidence of superordinacy/subordinacy is likely to be replaced by evidence of convergence/divergence. One may generally expect to find that constructs vary in their range of application and that there are several that overlap at any particular focus of interest.

To conclude, the theory that construct systems are hierarchical appears questionable to some extent. Ordinal terms can generally be avoided in discussing the relationships between constructs. Laddering, however, and probing in general, are procedures that may be employed with advantage in grid technique, even by people who prefer to adopt a non-commital attitude towards the theory and its speculative elaborations.

3.12 Missing data

The omission of a single entry creates a large gap in a grid. The complete row of entries for a construct defines a single point, its location in the element-space; if one entry in the row is missing its location there is unknown. Similarly the location of the element in the construct-space is only known if the column of entries referring to it is complete. For the purpose of principal component analysis either one element or one construct is lost if one entry in the grid is missing; all the data in one row or one column of it have to be erased. If he wishes to use this method of analysis the clinician is only left free to choose which can most easily be spared, the element or the construct.

A reasoned or inspired guess at the missing entry might perhaps be interpolated excusably if its omission could safely be assumed to have been accidental. but the assumption is extremely suspect. The possibility needs to be considered that the entry is missing because the element does not come within range of convenience of the construct.

Landfield discusses the psychological significance of missing entries in *Explorations* (Chapter 6) and proposes ways of using them for measurement. If eliminating the constructs and elements they affect does not leave a complete grid large enough to cover the region of interest, the best thing to do with the data may be to seek some explanation for the occurrence of the omissions.

The interviewer should take care to see that no entries are missing from his grid when he records it, and not try to 'correct' it later.

2.13 Conclusions

Altogether, constructing a grid for an individual informant is a comparatively simple operation. No complicated apparatus is needed. The interviewer does not have to rely on any psychological theory or accept any restrictive assumptions. If he feels satisfied that grid technique is worth applying in the case he is studying, he can go ahead and use it without troubling about how far it extends to other psychological problems.

Not for him are the labours of item analysis and standardization. But he must accept the risk of disappointment. Each grid designed for an individual is an attempt to explore a new unknown intrapersonal space. Some general ideas can be given about directions to follow and there is fairly detailed advice on methods of measurement, but what will be discovered is unpredictable. The proof of the grid is the results obtained from it — not from grids constructed by other interviewers for different informants on other occasions.

It needs years of study, consulting ponderous textbooks and turning up abstruse references, understanding theories and mastering methods, to appreciate precisely how much there is in academic psychology that can be safely, gratefully ignored in using grid technique. What the interviewer needs most are open-minded curiosity, respect for evidence, good rapport and sympathetic understanding.

REFERENCES TO PART I

Bannister, D., and Mair, J. M. M. (1968). *The Evaluation of Personal Constructs*, Academic Press, London.

Blake, W. (1925). 'The marriage of heaven and hell'. In G. Keynes (Ed.), *The Writings of William Blake*, Nonesuch Press, London.

Breuer, J. (1955). 'Case history of Fraulien Anne Q'. In *The Complete Works of Sigmund Freud, 2*, Hogarth Press, London.

Burt, C. (1940). *Factors of the Mind*, University of London Press.

Dresser, I. G. (1969). *Repertory Grid Technique in the Assessment of Psychotherapy*, M. Phil. thesis, University of London.

Eysenck, H. J. (1957). *The Dynamics of Anxiety and Hysteria*, Routledge and Kegan Paul, London.

Feldman, M. M. (1972). *The Body Image Review and Exploration of an Experimental Approach*, M. Phil. thesis, University of London.

Galton, F. (1883). *Inquiries into Human Faculty*, Macmillan, London.

Gray, J. (1965). 'A time-sample of the components of general activity in selected strains of rats'. *Canadian J. Psychol., 19*, 74–82.

Hinkle, D. N. (1965). *The Change of Personal Constructs from the Viewpoint of a Theory of Implications*, Dissertation, Ohio State University.

Hinkle, D. N. (1970). 'The game of personal constructs'. In D. Bannister (Ed.), *Perspectives in Personal construct Theory*, Academic Press, London.

Holland, R. (1970). 'George Kelly, constructive innocent and reluctant existentialist'. In D. Bannister (Ed.), *Perspectives in Personal Constructs Theory*, Academic Press, London.

Kelly, G. A. (1955). *The Psychology of Personal Constructs*, Norton, New York.

Kelly, G. A. (1963). *A Theory of Personality*, Norton, New York.

Keynes, J. M. (1921). *A Treatise on Probability*, Macmillan, London; Reprinted in *The Collected Writings of John Maynard Keynes*, Macmillan, London and New York (1973).

Jung, C. G. (1957). *Collected Works*, Routledge and Kegan Paul, London.

Landfield, A. W. (1971). *Personal Construct Systems in Psychotherapy*, Rand McNally, Chicago.

McCully, R. S. (1971). *Rorschach Theory and Symbolism*, Williams and Watkins, Baltimore.

Moreno, J. L. (1934). *Who Shall Survive?*, Nervous and Mental Disease Pub. Co., Washington, D.C.

Neumann, E. (1954). *The Origins and History of Consciousness*, Routledge and Kegan Paul, London.

Neumann, E. (1955). *The Great Mother*, Routledge and Kegan Paul, London.

Oliver, D. (1970). 'George Kelly: an appreciation'. In D. Bannister (Ed.), *Perspectives in Personal Construct Theory*, Academic Press, London.

Orley, J., and Leff, J. (1972). 'The effect of psychiatric education on attitudes to illness among the Ganda'. *Brit. J. Psychiat., 121*, 137–141.

Osgood, C. E., Suci, G. J., and Tannenbaum, P. H. (1957). *The Measurement of Meaning*, University of Illinois Press, Urbana, Chicage and London.

Ryle, A., and Lunghi, M. (1970). 'The dyad grid: a modification of repertory grid technique'. *Brit. J. Psychiat.*, 117, 323–327.

Shapiro, M. B. (1961). 'A method of measuring psychological changes specific to the individual psychiatric patient'. *Brit. J. med. Psychol.*, 34, 151–155.

Slater, P. (1960). *The Reliability of Some Methods of Multiple Comparison in Psychological Experiments*, Ph. D. thesis, University of London.

Slater, P. (1965). 'The test–retest reliability of some methods of multiple comparison'. *Brit. J. mat. and stat. Psychol.*, 18, 227–242.

Stagner, R., and Osgood, C. E. (1946). 'Impact of war on a nationalistic frame of reference: 1, Changes in general approval and qualitative patterning of certain stereotypes'. *J. soc. Psychol.*, 46, 187–215.

Stephenson, W. (1935). 'Correlating persons instead of tests'. *Character and Personality*, 4, 17–24.

Stephenson, W. (1953). *The Study of Behaviour: Q-technique and Its Methodology*, University of Chicago Press.

Szondi, L., Moser, V., and Webb, M. (1959). *The Szondi Test in Diagnosis, Prognosis and Treatment*, J. B. Lippicott, Philadelphia.

Thomson, G. (1948). *The Factorial Analysis of Human Ability*, University of London Press.

Titchener, E. B. (1912). *A Textbook of Psychology*, Macmillan, New York.

Topçu, S. (1976). *Psychological Concomitants of Aggressive Feelings and Behaviour*, Ph. D. thesis, University of London.

Torgerson, W. S. (1958). *Theory and Methods of Scaling*, John Wiley and Sons, London and New York.

Watson, J. B. (1913). 'Psychology as the behaviourist views it'. *Psych. Rev.*, 20, 158–177.

Watson, J. B. (1931). *Behaviourism*, Kegan Paul, Trench, Tribner and Co., London.

Wundt, W. (1911). *Grundriss der Psychologie*, Engelmann, Leipzig.

Acknowledgement

Chapters in this Part and Part IV include excerpts from two papers published in the *British Journal of Psychiatry*: 'The use of the repertory grid technique in the individual case' (1965, 111, 965–975), and 'Theory and technique of the repertory grid' (1969, 115, 1287–1296). The author thanks the editor for permission to incorporate them.

PART II

ANALYSIS

4

PROLOGUE

A clinician should consider that he is committing himself to a small-scale research project when he decides to use grid technique in the study of an individual patient. Although it can be constructed and recorded in one clinical session, or two at most, a single grid may contain as much data as a postgraduate student might not long ago, have collected in the course of a research project for a doctorate. The complexity of the relationships it reveals may well be greater.

Much incidental information may be gained in the course of choosing the terms for the grid and filling it in, and much may be extracted without any elaborate analysis. The patient may collaborate with the clinician at every stage. He may gain insight from finding himself evaluating elements in terms of constructs he had not thought of extending to them before, and communicate his insight to the clinician. If the grid has been filled in one construct at a time it may be reviewed one element at a time to study the complete picture of each element in terms of all the constructs. In this way fresh points of interest may be discovered.

A few simple computations may provide further information. In examining the addict's grid (Table 3.1), someone who has never felt blocked or experienced a good buzz personally may feel puzzled about what the expressions mean. If so, he may gather some ideas from the correlations between the constructs. Feeling blocked, for instance:

Correlates	With	Construct
0.09	Wanting to talk more	1
1.00	Feeling high	2
−0.11	Feeling sleepy	4
0.08	Having a warm feeling inside	5
−0.30	Feeling drunk	6
−0.02	Imagining things	7
−0.23	Feeling sick	8
0.29	Doing things without knowing	9
0.64	Enjoying things	10
0.46	Having a good buzz	11
−0.04	Feeling tense	12
−0.47	Feeling sexy	13
0.42	Seeing or hearing people who aren't there	14

The doctor attempting to cure his patient of addiction to heroin might attach more importance to the relative distances between it and the other drugs in his patient's construct spaces:

Distance	From
0.82	Alcohol
0.60	Barbiturates
0.74	Cannabis
1.12	Cocaine
1.27	Drynomil
0.94	L.S.D.
0.53	Mandrax
1.13	Methedrine injections

Note: standard distance 1.0

Though previous experience and clinical impressions might already have led him to expect that mandrax would alleviate withdrawal symptoms, he might be encouraged by finding that his patient reports it most similar in its effects. He might also be glad to have the relationships quantitatively defined. More information can be extracted by further analyses.

In pre-computer days the research student could devote a month or more to such a task and at the end might feel his time well spent. A complete analysis of a grid, such as that described in Chapter 9, would certainly take no less; even calculating the results just quoted might take about half a day. But the days of do-it-yourself computation are over. The clinician with easy access to a computer should not find it difficult to get his data accepted one day and his output returned the next — provided everything goes according to plan. The actual time needed for processing the data in a powerful computer is less than a second.

The clinician primarily interested in his patient does not really have to know how the computations are performed or why; all he needs to know is how to decipher and interpret his results. In short, he can afford to skip Chapters 5 and 6 and resume reading at Chapter 7. So, too, can those who already have a good knowledge of mathematics, including principal component analysis — the mathematics of eigenvalues. There is nothing in the two chapters not already well known to the expert. They are included for those who have come fresh to the subject and would like to know why a particular method is recommended for analysing a grid and also how it works

The analysis has nothing directly to do with the confirmation or refutation of hypotheses; it is only concerned with thorough examination of the observations.

5

A BRIEF INTRODUCTION TO MATRIX ALGEBRA AND COORDINATE GEOMETRY

5.1 Introduction

Students who do not pass beyond the ordinary level in mathematics — among whom are many who go on into medicine, psychology and the social sciences — may not be introduced to matrix algebra or coordinate geometry during any of their courses. Later, if they decide that some knowledge of the subjects may help them to solve problems that have come to interest them, they may run into difficulties trying to obtain it from some textbook of advanced mathematics, where it is combined with information on other unfamiliar subjects; and they may either give up in bewilderment or labour through a much more extensive course of study than is necessary for the purpose they have in mind. This chapter is intended to help such people over their difficulties, particularly those whose curiosity derives from their interest in the results that can be obtained from grids. Grids, being tables with a row for every construct and a column for every element, are two-way arrays which come under the comprehensive term 'matrices'. Matrix algebra is needed to explain how they can be analysed. It applies mathematics to whole arrays, treating them by methods corresponding to the ones used for single quantities in elementary algebra. This chapter explains how the operations of addition, subtraction, multiplication and division are carried out, so that familiarity with them may be safely assumed later. Some other accessory operations which have no parallel in elementary mathematics, such as transposition, are also explained.

The entries in a grid have geometrical properties as well, because of which, for instance, distances between elements in an informant's intrapersonal space can be measured and compared. A knowledge of coordinate geometry sufficient for understanding these properties may make grids more illuminating and easier to interpret. Some information on the subject has accordingly been introduced at the end of the chapter.

5.2 Notation

When a single symbol may indicate a single quantity or an array of any size, notation which will avoid confusion needs to be chosen carefully. Distinctions can be made by using different kinds of lettering. The custom followed here is to use italic type for single values (scalars) and bold-face type for arrays, with two-way arrays (matrices) in capitals and one-way arrays (vectors) in lower-case lettering. For example, if a variable, say blood pressure, were measured in one case on one occasion the reading might be denoted x; if it were measured in the same case on a series of occasions the readings could be listed as a row of entries and the whole list would be correspondingly denoted x; if it were measured in several cases on a series of occasions the readings could be recorded in a table with a row for each case and a column for each occasion, and the whole table would be denoted **X**.

Subscripts are also useful for making statements precise and clear. In general, suppose a grid recording the evaluations of m elements in terms of n constructs is tabulated in an array with n rows and m columns, we may use i to indicate one particular number in the range from 1 to n and j for one in the range from 1 to m. Then g_{ij} will indicate one particular entry in the grid, namely the one in row i column j which records the evaluation of element j in terms of construct i.

When subscripts are combined in pairs like this the convention has been established that the first refers to a row and the second to a column in the matrix. Accordingly, the array **M**, which appears later in this section, can be described as 4 by 3 because it has 4 rows and 3 columns, while its transpose **M**$'$, which appears in the next section, is 3 by 4.

The convention is adopted here that grids are listed construct by element, and in reference to them the number of constructs is denoted n and the number of elements m. So any **G** is n by m unless otherwise stated. When referring to a series of grids with matching elements and constructs, o is used for the number in the series. In other contexts the Fortran convention is followed in that the letters i to n denote integers, and when grids are not specifically concerned m and n are used unrestrictedly for any integral number.

A convenient use has recently been adopted where the dot or full stop in subscript is used to indicate a complete range of values. For instance, if we want to refer to all the entries in a given row of the grid, say row i, we may denote them by the vector $g_{i.}$, where the dot indicates that the values of j range from 1 to m. Similarly we may use $g_{.j}$ to denote all the entries in column j, for $i = 1, \ldots, n$.

The sum of all the terms in a vector will be indicated by printing the letter that refers to it as a capital in italics and their mean by the same letter in lower-case italics. The dots in subscript indicate the set of entries over which summation or averaging extends. Thus the sum of all the entries in the vector $g_{i.}$ is $G_{i.}$ and their mean is $g_{i.}$. Similarly the sum of all the entries in the matrix **G** is $G_{..}$ and their mean is $g_{..}$. Sums of other terms will be denoted when necessary by the symbol Σ. For example, the expression

$$\Sigma(g_{ij} - g_{i.})^2 \qquad (j = 1, \ldots, m)$$

could by used to denote the sum of the squares of the differences between the entries in row i and their mean.

Square brackets have long been used to define the contents of a matrix. For example, the equation

$$M = \begin{bmatrix} 5 & 0 & 3 \\ -4 & -2 & 1 \\ 0 & -3 & 4 \\ 2 & -1 & -5 \end{bmatrix}$$

gives the contents of a particular matrix, M. A grid might be written out in an extended form as

$$\begin{bmatrix} g_{11} & \cdots & g_{1j} & \cdots & g_{1m} \\ \cdots & \cdots & \cdots & \cdots & \cdots \\ g_{i1} & \cdots & g_{ij} & \cdots & g_{im} \\ \cdots & \cdots & \cdots & \cdots & \cdots \\ g_{n1} & \cdots & g_{nj} & \cdots & g_{nm} \end{bmatrix}$$

or again as

$$[g_{ij}] \quad (i = 1, \ldots, n; \quad j = 1, \ldots, m)$$

to indicate that it contains all the values of g_{ij} obtained by allowing i to range from 1 to n and j from 1 to m. When g_{ij} is used like this it is called the characteristic entry in G. If the numbers of rows and columns in a matrix need special attention they may be written in subscript to left and right of the matrix, which is bracketed, thus:

$$G = {}_n[G]_m$$

5.3 Transposition

Transposing a matrix is simply converting its rows into columns: the entries in row i of the matrix become the entries in column i of its transpose. A prime is used to indicate that this operation has been carried out. For instance, the transpose of M, listed above, is

$$M' = \begin{bmatrix} 5 & -4 & 0 & 2 \\ 0 & -2 & -3 & -1 \\ 3 & 1 & 4 & -5 \end{bmatrix}$$

Thus if the blood pressures previously mentioned had been tabulated with a row for each occasion and a column for each case, the resulting table would have been X' instead of X.

The distinction between a matrix and its transpose is vital. If a computer were instructed to read M as a 4 by 3 matrix, then given M' the matrix that would pass on to the next stage in the processing would be

$$\begin{bmatrix} 5 & -4 & 0 \\ 2 & 0 & -2 \\ -3 & -1 & 3 \\ 1 & 4 & -5 \end{bmatrix}$$

and the eventual output might well be incorrect.

Transposition can be applied to vectors as well as matrices. When a row vector is transposed it becomes a column vector, and vice versa.

When a matrix which has already been transposed is transposed again it returns to its original form. This may be stated as $M'' = M$. The only kind of matrix which has the property $M' = M$, that is to say, which remains unchanged by transposition, is a symmetrical square matrix such as

$$\begin{bmatrix} a & b & c & d \\ b & e & f & g \\ c & f & h & i \\ d & g & i & j \end{bmatrix}$$

More precisely a square matrix, say S, is symmetrical if $s_{ij} = s_{ji}$ for all i and j.

The set of entries with $i = j$, running from the top left-hand corner to the bottom right, forms its leading diagonal, and the matrix is symmetrical about that.

A matrix cannot be symmetrical unless it is square, but there are any number of matrices which are square without being symmetrical. There are also certain symmetrical square matrices where every entry outside the leading diagonal is zero; they are called diagonal matrices.

5.4 The treatment of simultaneous linear equations by matrix algebra

Though matrix algebra usually makes its first appearance in advanced courses in mathematics its relevance reaches back to a far earlier stage — it applies to the solution of simultaneous linear equations. In his earliest course in algebra the pupil learns to solve them by elimination and substitution. For example, he might be set the problem (in old British currency and weight):

Mrs. Archer paid 17d for 3 lb of apples and 2 lb of plums.
Mrs. Baynes paid 22d for 2 lb of apples and 4 lb of plums.
They both paid the same price per pound for each kind of fruit.
What was it?

He may set about solving the problem by writing down the equations

$$3x + 2y = 17 \tag{5.1}$$

$$2x + 4y = 22 \tag{5.2}$$

where x and y stand for the unknown prices per pound. Then multiplying equation (5.1) by 2 and subtracting equation (5.2) eliminates y and leaves him with the equation $4x = 12$, showing that the apples were 3d per pound. Lastly, substituting 3 for x in equation (5.1) shows that the plums cost 4d per pound.

This working method may do to solve equations in two or three unknowns when the answers can be given in round numbers, but beyond these limits it soon becomes unmanageable. Matrix algebra provides a general procedure for solving such equations with any number of unknowns. Of course it is useful for many other purposes, too, but referring to this application helps to explain in the simplest way how it works.

Let us call the two unknown values x_1 and x_2, to allow scope for extending their number indefinitely, the amount required in the first case a_{11} and a_{12} respectively and the total cost c_1; and let us use a_{21}, a_{22} and c_2 similarly for the second case. Then we have a general model for a pair of equations with two unknowns:

$$a_{11}x_1 + a_{12}x_2 = c_1 \atop a_{21}x_1 + a_{22}x_2 = c_2 \Bigg\} \text{Set 1}$$

They can readily be extended to n equations for m unknowns:

$$\begin{aligned} a_{11}x_1 + \cdots + a_{1j}x_j + \cdots + a_{1m}x_m = c_1 \\ \cdots \quad \cdots \quad \cdots \quad \cdots \quad \cdots \quad \cdots \\ a_{i1}x_1 + \cdots + a_{ij}x_j + \cdots + a_{im}x_m = c_i \\ \cdots \quad \cdots \quad \cdots \quad \cdots \quad \cdots \quad \cdots \\ a_{n1}x_1 + \cdots + a_{nj}x_j + \cdots + a_{nm}x_m = c_n \end{aligned} \Bigg\} \text{Set 2}$$

The operations can all be represented as additions if the products, typically $a_{ij}x_j$, are free to be positive zero or negative.

The equations all have the same construction. Take the first in set 1. It is composed of two sets of terms which may be written as vectors: $[a_{11} \ a_{12}]$ and $[x_1 \ x_2]$. The scalar c_1 is obtained by summing the products of the successive terms in them. And every other equation can be analysed in the same way. We can write a general expression for them all:

$$\mathbf{a} : \mathbf{x} = c$$

where the vector a contains a set of coefficients and x the values to which they refer, the colon represents the operation of summing the products of the associated terms in the two vectors and the scalar c is its result.

The operation represented here by the colon will be examined more closely later. It is a form of multiplication.

5.5 Addition and Subtraction

The two equations in the schoolboy's problem can obviously be added together to make

$$5x + 6y = 39$$

which is another equation of the same form as the two before. The two may be written as

$$\mathbf{a}_1 : \mathbf{x} = c_1$$

$$\mathbf{a}_2 : \mathbf{x} = c_2$$

and their sum as

$$\mathbf{a}_s : \mathbf{x} = c_1 + c_2$$

Here

$$\mathbf{a}_1 = [3 \quad 2]$$

$$\mathbf{a}_2 = [2 \quad 4]$$

and

$$\mathbf{a}_s = [5 \quad 6]$$

The successive terms in a_s are obtained by adding the corresponding terms in a_1 and a_2. They can be added because they are aligned to the successive values of x. This must be the correct rule to follow. The schoolboy would have got his answer wrong it he had confused the plums with the apples. What might have happened if such things as soap-flakes, sausages and cigarettes had been included on the shopping list as well?

Any number of vectors of the same length can be added provided they are aligned. Summing the coefficients of each x in the equations of set 2, which form the vectors

$$a_{.1} \quad \cdots \quad a_{.j} \quad \cdots \quad a_{.m}$$

yields a vector of the sums $a_{s.}$, with the characteristic entry

$$a_{sj} = a_{1j} + \cdots + a_{ij} + \cdots + a_{nj}$$

The same rule governs the addition of matrices; they can be added provided they are aligned by row and by column. Consider another set of equations referring to the same unknowns in the same cases as the ones in set 2, say:

$$b_{11}x_1 + \cdots + b_{1j}x_j + \cdots + b_{1m}x_m = d_1$$

$$\cdots \quad \cdots \quad \cdots \quad \cdots \quad \cdots \quad \cdots$$

$$b_{i1}x_1 + \cdots + b_{ij}x_j + \cdots + b_{im}x_m = d_i$$

$$\cdots \quad \cdots \quad \cdots \quad \cdots \quad \cdots \quad \cdots$$

$$b_{n1}x_1 + \cdots + b_{nj}x_j + \cdots + b_{nm}x_m = d_n$$

The vectors

$$c = [c_1 \quad \cdots \quad c_i \quad \cdots \quad c_m] \quad \text{and} \quad d = [d_1 \quad \cdots \quad d_i \quad \cdots \quad d_m]$$

can be added because they are aligned by case. Their sum will be a vector of m terms, with the characteristic entry

$$c_i + d_i = (a_{i1} + b_{i1})x_1 + \cdots + (a_{ij} + b_{ij})x_j + \cdots$$

Consequently the two arrays of coefficients

$$A, \text{ i.e. } [a_{ij}] \quad \text{and} \quad B, \text{ i.e. } [b_{ij}]$$

can be added to form another n by m matrix, say S, where the characteristic entry is

$$s_{ij} = a_{ij} + b_{ij}$$

Subtraction does not present any further problem. It follows the same rule as addition. Given two equations from set 2, for instance, it is just as simple to subtract one from the other as to add the two together. Thus

$$c_1 - c_2 = (a_{11} - a_{21})x_1 + \cdots + (a_{1j} - a_{2j})x_j + \cdots$$

Accordingly, the differences between the two aligned vectors $a_{1.}$ and $a_{2.}$ form a vector, also of m terms, referring to the successive values of x; its characteristic entry is $a_{1j} - a_{2j}$. Therefore the same reasoning as before leads to the conclusion that one two-way array can be subtracted from another provided they are aligned by row and by column. The differences between A and B, as defined above, form an n by m matrix, say D, with the characteristic entry

$$d_{ij} = a_{ij} - b_{ij}$$

Since any matrix can be added to itself any number of times it can be multiplied by any scalar, say p. The effect will be to multiply every entry in it by p, i.e.

$$p[a_{ij}] = [p.a_{ij}]$$

5.6 Multiplication

In matrix notation the two equations in set 1 can be condensed into the single expression

$$A : x = c$$

where A is the 2 by 2 array

$$\begin{bmatrix} a_{11} & a_{12} \\ a_{21} & a_{22} \end{bmatrix}$$

The vectors x and c contain two entries each, and the colon denotes the operation that combines A and x to produce c. The same expression will serve to summarize all the equations in set 2 if A is redefined as an n by m array, each row of which contains a set of coefficients of the m values of x in x, and c is the set of n values of c obtained by combining A with x. The operation which combines them is now due for closer examination.

Let us take the two equations in set 1:

$$a_{11}x_1 + a_{12}x_2 = c_1$$
$$a_{21}x_1 + a_{22}x_2 = c_2$$

A vector, say $p = [p_1 \ p_2]$ may be introduced to combine with c in just the same way as the terms in the first row of A combined with x in the equation for c_1. A scalar quantity q will be obtained which is the sum of the products $p_1 c_1$ and $p_2 c_2$. This can be stated in elementary terms as $q = p_1 c_1 + p_2 c_2$. In matrix notation, using the colon, it can be stated as $q = p : c$. Also, as $c = A : x$, it can be stated as $q = p : A : x$.

Elementary algebra can be used to calculate q from p, A and x. Then the results can be examined to discover precisely what the expression $p : A$ signifies. We start with

$$q = p_1 c_1 + p_2 c_2$$
$$= p_1(a_{11}x_1 + a_{12}x_2) + p_2(a_{21}x_1 + a_{22}x_2)$$

Removing the brackets and rearranging the terms gives us

$$q = (p_1 a_{11} + p_2 a_{21})x_1 + (p_1 a_{12} + p_2 a_{22})x_2$$

So a vector of two terms

$$[p_1 a_{11} + p_2 a_{21} \quad p_1 a_{12} + p_2 a_{22}]$$

is the resultant of $\mathbf{p} : \mathbf{A}$ and is described as the product of premultiplying $\mathbf{A}$ by $\mathbf{p}$. The operation and its result may be written out in full as

$$[p_1 p_2]\begin{bmatrix} a_{11} & a_{12} \\ a_{21} & a_{22} \end{bmatrix} = [p_1 a_{11} + p_2 a_{21} \quad p_1 a_{12} + p_2 a_{22}]$$

The first term in the product is the sum obtained by adding the products of the terms in $\mathbf{p}$, a *row* vector, with the terms in the first *column* of $\mathbf{A}$; and the second is the sum of their products with the terms in the second column of $\mathbf{A}$. This establishes the rule of multiplication in matrix algebra, namely *row into column*, which is always rather a surprising one to the newcomer.

The example can easily be extended further. We can introduce a second pair of coefficients and a third, or as many more as we like, say k altogether; combining them with $\mathbf{c}$ will produce k values of $\mathbf{q}$, i.e.

$$\mathbf{P} = \begin{bmatrix} p_{11} & p_{12} \\ p_{21} & p_{22} \\ \cdots & \cdots \\ p_{k1} & p_{k2} \end{bmatrix}$$

multiplied into $\mathbf{Ax}$ gives

$$\begin{bmatrix} q_1 \\ q_2 \\ \cdots \\ q_k \end{bmatrix} = \mathbf{q}$$

To keep to this rule we must write $\mathbf{x}$ as a column vector when we write it out in full for the equation $\mathbf{Ax} = \mathbf{c}$. Thus set 1 would be written as

$$\begin{bmatrix} a_{11} & a_{22} \\ a_{21} & a_{22} \end{bmatrix}\begin{bmatrix} x_1 \\ x_2 \end{bmatrix} = \begin{bmatrix} c_1 \\ c_2 \end{bmatrix}$$

And the equation $\mathbf{PAx} = \mathbf{q}$, giving the values of $\mathbf{q}$ obtained by using k sets of coefficients in $\mathbf{P}$, would appear in its expanded form as

$$\begin{bmatrix} p_{11} & p_{12} \\ p_{21} & p_{22} \\ \cdots & \cdots \\ p_{k1} & p_{k2} \end{bmatrix}\begin{bmatrix} a_{11} & a_{12} \\ a_{21} & a_{22} \end{bmatrix}\begin{bmatrix} x_1 \\ x_2 \end{bmatrix} = \begin{bmatrix} q_1 \\ q_2 \\ \cdots \\ q_k \end{bmatrix}$$

If we wanted to apply the same operation to the larger array in set 2, we would need n terms in each row of $\mathbf{P}$ to apply to the n terms in each column of $\mathbf{A}$ in accordance with the rule 'row into column'. Consequently $\mathbf{P}$ would become a k by n array. The operation $\mathbf{PAx}$ would produce k values of $\mathbf{q}$ as before. Using subscripts to indicate the number of rows and columns in each array makes this clear, viz:

$$_k[\mathbf{P}]_n \ _n[\mathbf{A}]_{mm}[\mathbf{x}]_1 = {_k[\mathbf{q}]_1}$$

The row into column rule implies that two matrices cannot be multiplied together unless the number of columns in the first equals the number of rows in the second. When this condition is satisfied the resultant has as many rows as the first and as many columns as the second.

A statement of the rule of matrix multiplication which is quite explicit and may be found useful in times of doubt is: The entry of row i column j of AB is the sum of the products of the entries in row i of A with the entries in column j of B. For example, to prove that the product of a matrix with its transpose is symmetrical (see the next section) it is enough to show that the entry in row i column j of the product is the same as the entry in row j column i.

5.7 Some further notes on multiplication

When two matrices are multiplied the second is sometimes said to be premultiplied by the first and the first postmulitplied by the second. The distinction does not imply any change in the rule of multiplication but is still worth noting. Any matrix can be premultiplied or postmultiplied by its transpose, and in general the results will be different: if B is n by m, premultiplication by its transpose produces $B'B$, which is m by m, and postmultiplication produces BB', which is n by n. The product is symmetrical in either case. For instance, applying the operations to the matrix M listed above produces

$$M'M = \begin{bmatrix} 45 & 6 & 1 \\ 6 & 14 & -9 \\ 1 & -9 & 51 \end{bmatrix} \text{ and } MM' = \begin{bmatrix} 34 & -17 & 12 & -5 \\ -17 & 21 & 10 & -11 \\ 12 & 10 & 25 & -17 \\ -5 & -11 & -17 & 30 \end{bmatrix}$$

A square matrix, say T, can be premultiplied or postmultiplied by itself and the result of both operations will be the same. This is really just a tautology as it amounts to saying that TT = TT. The product need not be symmetrical if T is not. What can be done once can be done again; the series of multiplications TTT... can be continued indefinitely. The successive powers of T so obtained may be denoted T^2, T^3, etc.

The next problem is to reverse this process and work back from T(or T^1) to T^0, T^{-1}, etc. For T^0 we need a matrix which will satisfy the equation

$$TT^0 = T^0 T = T$$

As can easily be verified the array which does so is a diagonal matrix, one with units in its leading diagonal and zeros elsewhere:

$$\begin{bmatrix} 1 & 0 & \ldots & 0 \\ 0 & 1 & \ldots & 0 \\ \ldots & \ldots & \ldots & \ldots \\ 0 & 0 & \ldots & 1 \end{bmatrix}$$

It must have the same number of rows and columns as T. It is called the identity matrix and denoted I. I can be applied to a matrix which is not square, but then its size must be varied to fit the needs of pre- or postmultiplication.

Beyond this is the problem of finding T^{-1}, the inverse or reciprocal of T, which must have the property

$$TT^{-1} = T^{-1}T = I$$

The solution provides matrix algebra with a method of division. It can be applied to simultaneous equations. Premultiplying both sides of the equation

$$Ax = c$$

by A^{-1} solves it for the vector x, for

$$A^{-1}Ax = Ix = x = A^{-1}c$$

The general method of inversion is not reached in one easy step, but the intermediate stages on the way to it are sufficiently important to make the entire argument worth following. It depends on the use of determinants, minors, cofactors and singularities, all of which need to be explained.

5.8 Determinants

To solve the pair of equations in set 1,

$$a_{11}x_1 + a_{12}x_2 = c_1 \tag{5.3}$$

$$a_{21}x_1 + a_{22}x_2 = c_2 \tag{5.4}$$

by elimination we might begin by multiplying equation (5.3) by a_{22} and equation (5.4) by a_{12} to produce

$$a_{11}a_{22}x_1 + a_{12}a_{22}x_2 = a_{22}c_1$$

and

$$a_{12}a_{21}x_1 + a_{12}a_{22}x_2 = a_{12}c_2$$

Then substraction eliminates x_2 and leaves

$$(a_{11}a_{22} - a_{12}a_{21})x_1 = a_{22}c_1 - a_{12}c_2$$

so

$$x_1 = \frac{a_{22}c_1 - a_{12}c_2}{a_{11}a_{22} - a_{12}a_{21}}$$

Similarly, multiplying equation (5.3) by a_{21} and equation (5.4) by a_{11} and then subtracting eliminates x_1 and leaves

$$x_2 = \frac{a_{21}c_1 - a_{11}c_2}{a_{12}a_{21} - a_{11}a_{22}}$$

The enumerators and denominators of these expressions have a general resemblance. The denominator in the equation for x_1 is the difference between the cross-products of the terms in A, i.e.

$$\begin{bmatrix} a_{11} & a_{12} \\ a_{21} & a_{22} \end{bmatrix}$$

This quantity is called its determinant and is usually denoted by enclosing the matrix in vertical brackets:

$$|A| = \begin{vmatrix} a_{11} & a_{12} \\ a_{21} & a_{22} \end{vmatrix} = a_{11}a_{22} - a_{12}a_{21}$$

The enumerator in the equation for x_1 is the determinant of the array

$$\begin{bmatrix} c_1 & a_{12} \\ c_2 & a_{22} \end{bmatrix}$$

formed by inserting c_1 and c_2 in place of the coefficients of x_1 in A. Again, inserting c_1 and c_2 in place of the coefficients of x_2 in A forms the matrix

$$\begin{bmatrix} a_{11} & c_1 \\ a_{21} & c_2 \end{bmatrix}$$

with the determinant $a_{11}c_2 - a_{21}c_1$; and dividing it by $|A|$ gives the value of x_2 expressed as

$$\frac{a_{11}c_2 - a_{21}c_1}{a_{11}a_{22} - a_{12}a_{21}}$$

The schoolboy's problem, for instance, can be solved by putting

$$x = \frac{\begin{vmatrix} 17 & 2 \\ 22 & 4 \end{vmatrix}}{\begin{vmatrix} 3 & 2 \\ 2 & 4 \end{vmatrix}} = \frac{68 - 44}{12 - 4} = 3$$

and

$$y = \frac{\begin{vmatrix} 3 & 17 \\ 2 & 22 \end{vmatrix}}{\begin{vmatrix} 3 & 2 \\ 2 & 4 \end{vmatrix}} = 4$$

The procedure generalizes, but in a way that is not easy to operate. The determinant of a 3 by 3 array is a grizzly-looking expression containing six terms, each a multiple of three of the entries in the array. And this is just a baby compared with the monstrous determinant of a 4 by 4 array, which contains four times as many terms, each a multiple of four entries. What next?

Once again, neat notation helps to make the nature of the operation clearer. The determinant of a 3 by 3 array can be written as

$$|\,A\,| = \begin{vmatrix} a_{11} & a_{12} & a_{13} \\ a_{21} & a_{22} & a_{23} \\ a_{31} & a_{32} & a_{33} \end{vmatrix}$$

$$= a_{11}(a_{22}a_{33} - a_{23}a_{32}) - a_{21}(a_{12}a_{33} - a_{13}a_{32}) + a_{31}(a_{12}a_{23} - a_{13}a_{22})$$

$$= a_{11}\begin{vmatrix} a_{22} & a_{23} \\ a_{32} & a_{33} \end{vmatrix} - a_{21}\begin{vmatrix} a_{12} & a_{13} \\ a_{32} & a_{33} \end{vmatrix} + a_{31}\begin{vmatrix} a_{12} & a_{13} \\ a_{22} & a_{23} \end{vmatrix}$$

That is to say, it is the sum of the multiples of the determinants of three 2 by 2 arrays. The first entry a_{11}, is multiplied by the determinant of its minor, i.e. the remaining part of A after the entries in the row and in the column containing a_{11} have been removed from it. The next entry, continuing down column 1, is a_{21}; it, too, is multiplied by the determinant of its minor. And lastly, a_{31} is used in the same way.

The determinantal coefficient of the first term, a_{11}, is positive; the coefficient of the next, a_{21}, is negative; and that of the third, a_{31}, is positive again. A formula is needed to define these alternations in sign. Referring to any entry in the array, say a_{ij}, we may use A_{ij} to denote its minor, i.e. the array derived from A by omitting all the entries in row i and all in column j, and then define its cofactor, A_{ij} which is the determinant of the minor including the appropriate sign, as

$$A_{ij} = (-1)^{i+j}|\,A_{ij}\,|$$

With this definition the determinant of the 3 by 3 array given above can be reexpressed as

$$|\,A\,| = a_{11}A_{11} + a_{12}A_{12} + a_{13}A_{13}$$

The solutions of the simultaneous equations would not be affected if they were written in a different order or if the order of the unknowns in them were changed: the schoolboy, for instance, could still get the right answer if he wrote down the equation for Mrs. Baynes's shopping before Mrs. Archer's or used x to refer to the plums instead of the apples. Accordingly, the order of the rows and the columns in a matrix is immaterial when a determinant is calculated by this method; we can follow any row or column we like. We would arrive at the same result for the 3 by 3 array above by following row 3 instead of column 1 and calculating its determinant as

$$|\,A\,| = a_{31}A_{31} + a_{32}A_{32} + a_{33}A_{33}$$

The general rule for this method, which is called expansion, is

$$|\,A\,| = \Sigma a_{ij}A_{ij}$$

where the summation extends to all the entries in any one row, i, or column, j, of A. Since it can be calculated in either way the determinant of any square matrix and of its transpose is the same.

5.9 Singularities

One generally expects to be able to find the solutions for n unknowns given n simultaneous equations, but it is not always possible. For example, the equations

$$3x + 2y - z = 2$$
$$2x - y + z = 5 \tag{5.5}$$
$$x - 4y + 3z = 8$$

cannot be solved. The reason is that one of them can be derived from the other two and does not provide any independent information. Any such lack of independence, which is called a singularity, produces a matrix with a zero determinant. Thus the determinantal ratio for x is

$$x = \frac{\begin{vmatrix} 2 & 2 & -1 \\ 5 & -1 & 1 \\ 8 & -4 & 3 \end{vmatrix}}{\begin{vmatrix} 3 & 2 & -1 \\ 2 & -1 & 1 \\ 1 & -4 & 3 \end{vmatrix}} = \frac{0}{0}$$

and the ratios for y and z are similar.

It is impossible to say which of the three equations is the dependent one. The singularity implies that any of them can be derived from the other two. Let a, b and c be the three equations of (5.5) Then

$$a = 2b - c$$
$$b = a + c/2$$

and

$$c = 2b - a$$

Though it may look as if there are three, in effect there are only two, which is not enough for finding three unknowns.

5.10 Pivotal condensation

Singularities can be used to simplify the calculation of determinants. Suppose we add a vector b to one of the columns in **A** — say it first column — to form the matrix

$$\mathbf{T} = \begin{matrix} (a_{11} + b_1) & a_{12} & \cdots & a_{1n} \\ (a_{21} + b_2) & a_{22} & \cdots & a_{2n} \\ \cdots \cdots & \cdots \cdots & \cdots \\ (a_{n1} + b_n) & a_{n2} & \cdots & a_{nn} \end{matrix}$$

and then obtain its determinant by expansion down that column. The result will be

$$|T| = (a_{11} + b_{11})A_{11} + \cdots + (a_{n1} + b_{n1})A_{n1}$$

$$= (a_{11}A_{11} + \cdots + a_{n1}A_{n1}) + (b_1 A_{11} + \cdots + b_n A_{n1})$$

that is to say, $|T|$ is the sum of the determinants of A and the matrix

$$\begin{bmatrix} b_1 & a_{12} & \cdots & a_{1n} \\ b_2 & a_{22} & \cdots & a_{2n} \\ \cdots & \cdots & \cdots & \cdots \\ b_2 & a_{n2} & \cdots & a_{nn} \end{bmatrix} = Q, \text{ say}$$

where b has replaced $a_{.1}$ in A.

Now if b has been taken from any of the other columns in A the determinant of Q will be zero, leaving $|T| = |A|$.

Thus in general $|A|$ can be obtained by summing $(a_{ij} - b_i) A_{ij}$ instead of $a_{ij}A_{ij}$ over row i, provided b is linearly dependent on any of the other rows in A or, alternatively, by summing $(a_{ij} - b_j) A_{ij}$ over column j, provided b is linearly dependent on any of the other columns.

This is the rationale of the procedure for calculating determinants by pivotal condensation. An example may help to explain how it works, starting with

$$A = \begin{bmatrix} 2 & 4 & 5 & -1 \\ -4 & 2 & -6 & 3 \\ 1 & 7 & 7 & 3 \\ 3 & 1 & 6 & 5 \end{bmatrix}$$

Multiples of $a_{1.}$ are added to the lower rows to reduce their entries in the first column to zero. Thus A is replaced by

$$B = \begin{bmatrix} 2 & 4 & 5 & -1 \\ 0 & 10 & 4 & 1 \\ 0 & 5 & 4.5 & 3.5 \\ 0 & -5 & -1.5 & 6.5 \end{bmatrix}$$

where

$$b_{1.} = a_{1.}$$

$$b_{2.} = a_{2.} + 2a_{1.}$$

$$b_{3.} = a_{3.} + \frac{a_{1.}}{2}$$

$$b_{4.} = a_{4.} - \frac{3a_{1.}}{2}$$

Similarly multiples of $b_{2.}$ are added to the lower rows of B to reduce their entries in the second column to zero, replacing it by

$$C = \begin{bmatrix} 2 & 4 & 5 & -1 \\ 0 & 10 & 4 & 1 \\ 0 & 0 & 2.5 & 3 \\ 0 & 0 & 0.5 & 7 \end{bmatrix}$$

where

$$c_{2.} = b_{2.}$$

$$c_{3.} = b_{3.} - \frac{b_{2.}}{2}$$

$$c_{4.} = b_{4.} + \frac{b_{2.}}{2}$$

As A is 4 by 4 the process is completed in one more operation of the same kind, which reduces C to

$$|D| = \begin{bmatrix} 2 & 4 & 5 & -1 \\ 0 & 10 & 4 & 1 \\ 0 & 0 & 2.5 & 3 \\ 0 & 0 & 0 & 6.4 \end{bmatrix}$$

where

$$d_{1.} = c_{1.}$$
$$d_{2.} = c_{2.}$$
$$d_{3.} = c_{3.}$$

$$d_{4.} = c_{4} - \frac{c_{3.}}{5}$$

This is a triangular matrix with all the entries below its leading diagonal reduced to zero, and its determinant is simply the product of the terms in its leading diagonal. For the general expression for $|D|$, namely

$$|D| = d_{11}D_{11} + d_{21}D_{21} + d_{31}D_{31} + d_{41}D_{41}$$

it reduces to

$$|D| = d_{11}D_{11}$$

because $d_{21} = d_{31} = d_{41} = 0$. Similarly,

$$D_{11} = 10 \begin{bmatrix} 2.5 & 3 \\ 0 & 6.4 \end{bmatrix}$$

since $d_{32} = d_{42} = 0$. Thus we arrive at

$$|A| = |B| = |C| = |D| = 2 \times 10 \times 2.5 \times 6.4 = 320$$

It is not necessary but it is usually simplest to take the terms in the leading diagonals of the successive arrays as the pivots. What is essential is to use singularities to replace the original square matrix by a triangular matrix with the same determinant.

Pivotal condensation has a very wide range of applications. It is used in solving multiple regression equations, in calculating discriminant functions and in multivariate analysis generally — in fact wherever simultaneous equations with large numbers of unknowns need to be solved.

5.11 Inversion

As every entry in an n by n matrix A has a corresponding cofactor A_{ij} the complete set of cofactors forms another n by n matrix with A_{ij} as its characteristic entry. If $[A_{ij}]$ is transposed and then multiplied by A an interesting result is obtained:

$$\begin{bmatrix} a_{11} & \cdots & a_{1i} & a_{1j} & \cdots \\ \cdots & \cdots & \cdots & \cdots & \cdots \\ a_{i1} & \cdots & a_{ii} & a_{ij} & \cdots \\ a_{j1} & \cdots & a_{ji} & a_{jj} & \cdots \\ \cdots & \cdots & \cdots & \cdots & \cdots \end{bmatrix} \begin{bmatrix} A_{11} & \cdots & A_{i1} & A_{j1} & \cdots \\ \cdots & \cdots & \cdots & \cdots & \cdots \\ A_{1j} & \cdots & A_{ij} & A_{ji} & \cdots \\ A_{1j} & \cdots & A_{ij} & A_{jj} & \cdots \\ \cdots & \cdots & \cdots & \cdots & \cdots \end{bmatrix} =$$

$$\begin{bmatrix} |A| & \cdots & 0 & 0 & \cdots \\ \cdots & \cdots & \cdots & \cdots & \cdots \\ 0 & \cdots & |A| & 0 & \cdots \\ 0 & \cdots & 0 & |A| & \cdots \\ \cdots & \cdots & \cdots & \cdots & \cdots \end{bmatrix}$$

Summing the products of the entries in row i of A and in column i of A_{ij} gives the determinant of A, while summing their products with the entries in any other column, j, results in zero. The reason is that any such sum of products forms the determinant of a singular matrix, where the entries in every row are the same as those in A except in row j, where the ones from row i appear in place of the originals. Consequently, the reciprocal of A is obtained by dividing the terms in the transposed matrix of its cofactors by its determinant, i.e.

$$\frac{[A_{ij}]'}{|A|} = A^{-1}$$

The matrix of cofactors for the 4 by 4 arrays used as an example in the previous section and its reciprocal (correct to four decimal places) are, respectively:

$$\begin{bmatrix} 331 & 137 & -180 & -10 \\ 127 & 69 & -100 & 30 \\ -245 & -55 & 140 & -10 \\ 137 & 19 & -60 & 50 \end{bmatrix} \text{ and } \begin{bmatrix} 1.0344 & .3969 & -.7656 & .4281 \\ .4281 & .2156 & -.1719 & .0594 \\ -.5625 & -.3215 & .4375 & -.1875 \\ -.0312 & .0938 & -.0312 & .1562 \end{bmatrix}$$

The statement

$$A^{-1} = \frac{[A_{ij}]'}{|A|}$$

is a precise mathematical definition of a reciprocal matrix and not a convenient algorithm for calculating one. There are computing programs for that. The formula shows that any square matrix can be inverted provided its determinant is not zero. Conversely no singular matrices (i.e. matrices containing any singularity) can be inverted because they must have zero determinants.

5.12 Coordinates

The connection between geometry and algebra is extremely close. Geometrical problems may be solved by algebra and algebraic ones by geometry. For instance, the schoolboy's problem with which this chapter started may be solved very easily by drawing a graph. Two straight lines can be drawn to represent the two equations and the point where they intersect will give the solution to the problem.

A surface is defined by drawing two axes, for x and y, at right angles away from a point of origin, O, and marking them off in units, (see Figure 5.1). All possible combinations of values of x and y which satisfy the equation

$$3x + 2y = 17$$

can be represented by a straight line on the surface. We may find two points on the line by substituting two arbitrary values for x and calculating the corresponding values of y; e.g. when $x = 1$, $y = 7$ and when $x = 5$, $y = 1$. The pair of values 1, 7 gives the location of one point, which is found by starting from the origin and moving 1 unit in the direction of x and then 7 units in the direction of y. It is marked as A1. The second, A2, is found by using the pair of values 5,1 in the same way. Every other point on the straight line passing through these two also satisfies the equation.

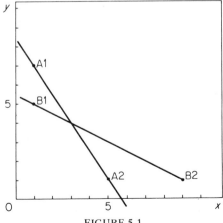

FIGURE 5.1

Similarly the straight line representing the second equation

$$2x + 4y = 22$$

must pass through the point B1, where $x = 1$ and $y = 5$; and through B2, where $x = 9$ and $y = 1$. The position of the point P, where the two lines intersect, can then be read from the graph as the one where $x = 3$ and $y = 4$. This is the only point which satisfies both equations and therefore answers the problem.

One number — a scalar — is sufficient to locate a point on one scale or axis; its sign shows the direction away from the origin and its magnitude shows the distance along the line. On a surface defined by two axes, such as x and y, a pair of numbers is needed to define the position of a point, such as A1 or P. The two together form a vector, that is to say, a set of directions for reaching the point from the origin. Three axes are needed to define a three-dimensional space, and a vector of three coordinates to locate the position of a point in it; and so on. A space or hyperspace of k dimensions must have k axes, and a vector of k coordinates is needed to locate one point in it. Conversely, any list of the measurements of k variables in a single case forms a vector defining the location on one point in a k-space.

A grid with n rows and m columns may be read either by row or by column. Read by row it contains n lists of m numbers, giving the locations of n points in an m-space; read by column it contains m lists, giving the locations of m points in an n-space. The first describes the dispersion of the constructs in the element-space; the second, the dispersion of the elements in the construct-space. When the method given in the next chapter is used for analysing a grid, the properties of these dispersions and the relationships between them are examined.

5.13 Polar coordinates

Suppose a, b and c (taken for convenience to be all positive quantities) are the measurements of three variables A, B and C in a particular case. We may set up a three-dimensional frame, as illustrated roughly in Figure 5.2 with the axis for A running from back to front, the axis for B from left to right and, for C, from bottom to top. To reach the point a, b, c we may trace a path from O to P1 by moving a units forward; then go on to P2 by making a right-angled turn and moving sideways for b units; and finally reach P3 by making another right-angled turn and moving upwards for c units.

With a different set of instructions P3 can be reached from O directly. The line to it leaves O at an angle of b degrees to the A axis and rises at an angle of v degrees from the surface defined by the axis of A and B; P3 lies at a distance of r units from O along it. The equations

$$\tan b = \frac{b}{a}$$

$$\tan v = \frac{c}{(a^2 + b^2)^{\frac{1}{2}}}$$

$$r = (a^2 + b^2 + c^2)^{\frac{1}{2}}$$

define b, v and r in terms of a, b and c. The vector $[a, b, c]$ gives the Cartesian coordinates of P3; b, v and r are its polar coordinates.

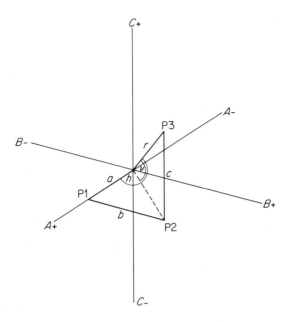

FIGURE 5.2

Cartesian coordinates are simplest for most purposes and apply to variation in any number of dimensions; polar coordinates have special advantages for describing variation in three dimensions. The advantages are greatest when the points to be mapped differ in their values of h and v but all have approximately equal values of r. In that case they diverge from O in different directions but lie at approximately the same distance away from it, i.e. on the surface of a sphere centered at O. If a three-dimensional dispersion is to be mapped on a globe of the kind intended for geographical purposes the locations of the points can be found easily from their polar coordinates; their Cartesian coordinates would be useless because it would be impossible to follow the tracks of their vectors through the interior of the globe.

6

THE RATIONALE OF PRINCIPAL COMPONENT ANALYSIS

6.1 The canonical form of an ellipse

Properties of ellipses are worth studying in connection with grids because the dispersions of elements in construct-spaces are elliptical in form. They are necessarily limited in every direction and tend to extend further in some than others. Even arrays of random numbers in the form of grids do not exhibit exactly equal amounts of variation in all directions; and in grids obtained from interviews with informants the tendency is much more marked.

People use interdependent constructs to reinforce one another, seeking confirmation of their judgements. Their construct systems consequently exhibit a great deal of redundancy. Perhaps because they see the outside world three-dimensionally they may succeed in adapting a three-dimensional frame of reference fairly readily for evaluating a set of elements within a common range of convenience, but generally they have difficulty in extending it further. Dispersions of elements in a construct-space are therefore always ellipsoidal and usually extremely so; they extend widely in a few directions, seldom more than three, and are closely contracted in the rest. Whatever the explanation, this is certainly the commonest and most conspicuous psychological phenomenon revealed by grid technique.

The canonical form of an ellipse, i.e. its simplest mathematical form, is given by the equation

$$\frac{x^2}{a^2} + \frac{y^2}{b^2} = 1$$

Figure 6.1 illustrates this. The ellipse has its centre at the origin, O, where the X and Y axes intersect. The coordinates x, y gives the position of a point P on its circumference. Assuming that a^2 is greater than b^2, P is furthest away from O when it is on the X axis, i.e. when $y = 0$. Then its radial distance from O (the length of the line OP) is a. As y increases x decreases; P moves round the perimeter and comes nearer to O until it reaches the Y axis, when its radial distance has diminished to b. Thereafter, as it continues on its course round O, its radial distance increases until it reaches the X axis once more. The X axis is the major axis of the ellipse and the Y axis its minor axis.

When $a = b$ the point P remains at the same distance from O throughout its course. The figure ceases to be an ellipse and becomes a circle with no major or minor axis. Thus every ellipse has a major and a minor axis. They are known as its

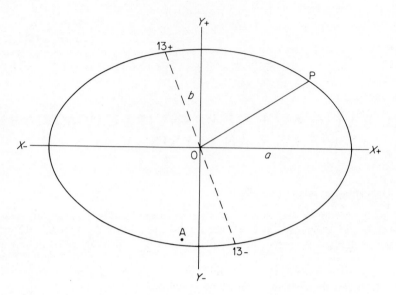

FIGURE 6.1

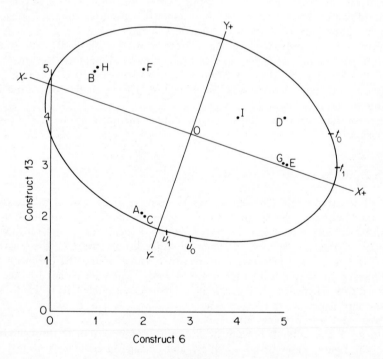

FIGURE 6.2

principal axes; they are at right angles to each other and the ellipse is symmetrical about them.

The equation for it becomes more complicated when it is not centered at the point of intersection of the reference axes (as in Figure 6.2, where the constructs are taken as the original reference axes) and its principal axes are not parallel to them. But if its centre and its principal axes are known, the original reference axes can be replaced by them and the ellipse can then be expressed in its canonical form.

A k-dimensional ellipsoid, similarly, has k principal axes at right angles to one another, about which it is symmetrical, and it is expressed in its simplest possible form when they are used as its reference axes.

6.2 Fitting an ellipse to a bivariate distribution

For a convenient example of this operation a miniature grid with $n = 2$ and $m = 9$ has been obtained by extracting the entries for constructs 6 and 13 from Table 3.1. They are repeated in Table 6.1. The distribution of the elements in the construct-space is mapped in Figure 6.2.

When an ellipse is to be fitted to it some rule is required for choosing the centre, the directions of the principal axes and the relative magnitudes of a and b. It must state exactly why they are preferable to any other parameters that might be proposed. The method of least squares gives the recognized rule for this purpose; it is one that has many other applications as well. It is explained in Section 6.4, after the values have been calculated.

The coordinates for the centre are given by the means (i.e. averages) of the constructs. The distances of the elements from this point, measured along the axes of the constructs, form another matrix, $\mathbf{D}$, also n by m, where the typical entry can be defined as

$$d_{ij} = g_{ij} - g_{i.}$$

In the example the means are 3 and 11/3, and the distances of the elements from them are shown in Table 6.2. A table of sums of squares and products

$$\mathbf{W} = \mathbf{DD}' \tag{6.1}$$

can be calculated from it.

In general, $\mathbf{W}$ is a symmetrical square matrix with n rows and columns. The entry in row i column i is the sum of squares of the entries in row i of $\mathbf{D}$, i.e. the total variation about the mean for construct i. It will be denoted V_i. The entry in row i column $j (j \neq i)$ is the total covariation between constructs i and j, and will be

TABLE 6.1

| Constructs | Elements | | | | | | | | | Total |
	A	B	C	D	E	F	G	H	I	
6	2	1	2	5	5	2	5	1	4	27
13	2	5	2	4	3	5	3	5	4	33

TABLE 6.2   Deviations from the construct mean

Constructs	Elements								
	A	B	C	D	E	F	G	H	I
6	−1	−2	−1	2	2	−1	2	−2	1
13	−5/3	4/3	−5/3	1/3	−2/3	4/3	−2/3	4/3	1/3

denoted W_{ij}. Accordingly W can be written out in detail as

$$
{}_n[W]_n =
\begin{matrix}
V_1 & \cdots & W_{1i} & W_{1j} & \cdots & W_{1n} \\
\cdots & \cdots & \cdots & \cdots & \cdots & \cdots \\
W_{i1} & \cdots & V_i & W_{ij} & \cdots & W_{in} \\
W_{j1} & \cdots & W_{ji} & V_j & \cdots & W_{jn} \\
\cdots & \cdots & \cdots & \cdots & \cdots & \cdots \\
W_{n1} & \cdots & W_{ni} & W_{nj} & \cdots & V_n
\end{matrix}
$$

Its symmetry is implied by $W_{ji} = W_{ij}$ for all i and j.

The total variation around the centre of the grid is the sum of the terms in the leading diagonal, namely

$$V = V_1 + \cdots + V_i + V_j + \cdots + V_n$$

W is called a covariance matrix and V its trace. In the example,

$$W = \begin{bmatrix} 24 & -5 \\ -5 & 12 \end{bmatrix}$$

and $V = 36$.

A diagonal matrix

$$
V =
\begin{bmatrix}
V_1 & 0 & \cdots & 0 \\
0 & V_2 & \cdots & 0 \\
\cdots & \cdots & \cdots & \cdots \\
0 & 0 & \cdots & V_n
\end{bmatrix}
$$

formed by replacing the terms outside the leading diagonal of W by zeros is useful in later computations (see Chapter 8).

The values of a^2 and b^2 needed for fitting the ellipse to the data in the example can be obtained by solving the determinantal equation

$$\begin{vmatrix} (24 - l) & -5 \\ -5 & (12 - l) \end{vmatrix} = 0$$

When expanded it becomes

$$l^2 - 36l + 263 = 0$$

and its roots, the values needed for a^2 and b^2, are found to be 25.81 and 10.19 approximately.

The general form of the determinantal (or characteristic) equation of W is

$$| W - l\mathbf{I} | = 0 \qquad (6.2)$$

When expanded it becomes a polynomial of order n, not easily solved without a computer. Its n roots will all be positive unless W contains singularities; then one or more of them will be zero. As this possibility cannot be excluded the number of non-zero roots will be denoted k, with the proviso $k \leqslant n$. The most convenient way of tabulating them is as a k by k matrix, with the successive roots from largest to smallest in its leading diagonal and zeros elsewhere. It will be denoted L.

If there are any zero latent roots the component-space does not extend into all the dimensions of the construct-space, but is confined to a k-dimensional subspace within it, the remaining $n - k$ dimensions being unoccupied by any of the elements. The complete set of k latent roots account for the whole of the observed variation, i.e.

$$V = l_1 + \cdots + l_k$$

Each component has a construct vector containing a coefficient for every construct to define its axis in the construct-space. So the complete set of these vectors forms a matrix

$$_n [C]_k$$

In general they are known as the latent vectors or eigenvectors of W. The equations defining the vector for component $i(i = 1 \ldots k)$ are

$$W c_{.i} = l_i c_{.i} \qquad (6.3)$$

and

$$c'_{.i} c_{.i} = 1 \qquad (6.4)$$

The first gives only the proportionate values of the coefficients; the second defines them precisely by requiring the vector to be normalized.

To obtain the vector for the major axis in the example the larger latent root is substituted for l_i in equation (6.3), which accordingly becomes

$$\begin{bmatrix} 24 & -5 \\ -5 & 12 \end{bmatrix} \begin{bmatrix} c_{11} \\ c_{21} \end{bmatrix} = \begin{bmatrix} 25.81 c_{11} \\ 25.81 c_{21} \end{bmatrix}$$

The two elementary equations obtained from it lead to the same result, namely that the proportionate values of the two coefficients are

$$\frac{c_{21}}{c_{11}} = -0.362$$

Equation (6.4) requires, furthermore, that

$$c_{11}^2 + c_{21}^2 = 1$$

Hence the normalized vector for the first component can be defined either as

$$\begin{bmatrix} c_{11} \\ c_{21} \end{bmatrix} = \begin{bmatrix} 0.94 \\ -0.34 \end{bmatrix}$$

or, in reverse as

$$\begin{bmatrix} c_{11} \\ c_{21} \end{bmatrix} = \begin{bmatrix} -0.94 \\ 0.34 \end{bmatrix}$$

The two pairs of coordinates indicate points on its axis lying opposite each other at unit distance from its origin, which is at the centre of the distribution. Both are equally appropriate because the axis, having its origin there, is necessarily bipolar. It appears in Figure 6.2 as $X\pm$.

The normalized vector for the second component can be defined similarly as

$$\begin{bmatrix} c_{12} \\ c_{22} \end{bmatrix} = \begin{bmatrix} 0.34 \\ 0.94 \end{bmatrix} \quad \text{or} \quad \begin{bmatrix} -0.34 \\ -0.94 \end{bmatrix}$$

and the axis for it appears as $Y\pm$.

The complete matrix of construct vectors can therefore be written as

$$C = \begin{bmatrix} 0.94 & 0.34 \\ -0.34 & 0.94 \end{bmatrix}$$

and it can be checked that pre- or postmultiplying C by its transpose yields an identity matrix. The variances of the vectors must equal 1 as they are normalized and their covariances must equal 0 as they are orthogonal (see Section 6.4).

In summary, the dispersion of the elements in the construct-space recorded by a grid is ellipsoidal. Its parameters are the vector of construct means, g_i., and the principal components of the covariance matrix W, which are specified by their latent roots L and their construct vectors C. In the example

$$g_i. = [3.0 \quad 3.6] \qquad\qquad W = \begin{bmatrix} 24 & -5 \\ -5 & 12 \end{bmatrix}$$

$$L = \begin{bmatrix} 25.81 & 0 \\ 0 & 10.19 \end{bmatrix} \quad \text{and} \quad C = \begin{bmatrix} 0.94 & 0.34 \\ -0.34 & 0.94 \end{bmatrix}$$

Figure 6.2 shows an ellipse fitted to the dispersion recorded in Table 6.1. Its centre is at g_i., its axes are orientated in the construct-space according to the directions in C and their relative lengths are proportional to the entries in the leading diagonal of $L^{\frac{1}{2}}$. The same ellipse has been shown in its canonical form in Figure 6.1.

6.3 The construct-space and the component-space

The main point of fitting the ellipse to the dispersion in Figure 6.2 is to show how the component-space is related to the construct-space. It is contained there and its location, extent and orientation are determined by the distribution of the elements there. Thus it is dependent on both the constructs and the elements.

For many purposes it may be sufficient to know the mathematical properties of the dispersion; there may be no need to draw an ellipse. If one is needed it will generally be to illustrate a particular aspect of a multidimensional ellipsoid, and the section likely to be of greatest interest will be the one with the two largest

TABLE 6.3   Element loadings (C'D)

Component	A	B	C	D	E	F	G	H	I
					Elements				
1	−0.37	−2.34	−0.37	1.77	2.11	−1.39	2.11	−2.34	0.83
2	−1.97	0.57	−1.91	1.00	0.05	0.91	0.05	0.57	0.65

components as its reference axes. It will cut across the axes of the constructs and will be shown most conveniently in its canonical form. If so, the elements will need to be plotted in their correct locations relative to the axes of the components, and the constructs in their correct orientations likewise.

The coordinates for the locations of the elements with reference to the axes of all the components can be obtained from $C'D$, and the ones for the orientations of the constructs can be read from C directly, by row. Those for the elements in the example are given in Table 6.3. They are described as loadings because they are weighted appropriately to allow for the difference in the total amount of variation along the axes of the two components. The weighting is expressed mathematically by the equation

$$C'DD'C = L \qquad\qquad\qquad (6.5)$$

To illustrate, element A has been plotted in Figure 6.1 from the coordinates given for it in Table 6.3. The orientation of the axis of construct 13 has also been indicated. The line for it connects the origin with the point defined by the vector $[-0.34\ 0.94]$, which has been taken from the second row of the example given for C in Section 6.2.

6.4 The method of least squares

There is, of course, some scope for differences of opinion about what is the best value to take, for instance, as the centre of a distribution. According to the method of least squares the best is the one that reduces the sum of the squares of the deviations to a minimum. For a series of m measurements of a single variable, X, this is their average or arithmetic mean, x. If any other value is taken instead, differing from x by d, the sum of the squares of the deviations from it will be $V + md^2$, where V is the sum of the squares of the deviations from x. The proof of this proposition does not introduce any assumptions about the form of the distribution of the measurements and is therefore valid whatever form it takes. Arguments starting from different assumptions about what is best converge remarkably often on the same conclusion.

The method of least squares has a wide range of applications. For example, it is applied systematically in the analysis of variance. It is mathematically convenient, the values which satisfy it being obtainable by differentiation.

How it applies to principal component analysis becomes clear when Hotelling's (1933) iterative method is used to calculate the latent roots and vectors of the components. His was the method most widely used before programs were developed for solving equation (6.2) directly (Wilkinson, 1965). It depends on equation (6.3), which can be simplified to

$$\mathbf{Wc} = l\mathbf{c}$$

for the first component. A trial vector, $\mathbf{t}_0$, containing 1 as its largest entry, is substituted for $\mathbf{c}$. Premultiplying it by $\mathbf{W}$ gives a vector whech can be written as $l_1 \mathbf{t}_1$. Its largest entry is an approximation to l, and $\mathbf{t}_1$, approximates closer to $\mathbf{c}$ than $\mathbf{t}_0$. The process is repeated with $\mathbf{t}_1$, giving

$$\mathbf{Wt}_1 = l_2 \mathbf{t}_2$$

and so on. Repeated iterations produce smaller and smaller changes in $\mathbf{t}$. When it stablizes, say after i iterations,

$$l_i = l \quad \text{and} \quad \mathbf{t}_i = \mathbf{c}$$

After the root and vector of the first component have been obtained a residual covariance matrix

$$\mathbf{W}_r = \mathbf{W} - l\mathbf{cc}'$$

is derived, from which specifications of the second can be calculated by the same method. If one latent root is not much larger then the next, convergence will be slow, but can be speeded up by powering the covariance matrix — forming $\mathbf{W}^2$, $\mathbf{W}^4$, $\mathbf{W}^8$, etc, — before iteration.

The process can be applied to the example of $\mathbf{W}$ which is reproduced in Table 6.4. Without referring to any of the other results already given one can tell, simply from looking at $\mathbf{W}$, that the major axis of the dispersion must follow the axis of construct 6 fairly closely, because that has the larger variation. As there is only a slight negative association between the two constructs it cannot diverge far towards the axis of 13. So one might do worse than start from the axis of 6, putting $\mathbf{t}_0' = [1 \ 0]$. Table 6.4 shows how the iteration proceeds from there. The values of $\mathbf{t}$ define successive positions of an axis OT. At its starting point it is parallel to the axis of construct 6, and it could be marked on Figure 6.2 by connecting O with the point $\mathbf{t}_0$ on the circumference of the ellipse. Successive iterations bring it closer to OX; the first brings it to $\mathbf{t}_1$.

As the space is only two-dimensional the simultaneous changes in an axis OU, at right angles to OT, can also be followed. It moves from U_0 (parallel to the axis of construct 13) through U_1 and so to OY as OT moves to OX.

When they are at $\mathbf{t}_0$ and $\mathbf{u}_0$ respectively their variation and covariation are of course those of the constructs, recorded in $\mathbf{W}$. The correlation between them, obtained by the usual formula

$$r_{ij} = \frac{W_{ij}}{(V_i V_j)^{1/2}}$$

comes to -0.295.

When they have been rotated to t_1 and u_1, their variation and covariation can be calculated by weighted summation. The normalized vectors for t_1 and u_1 are required. From the results in Table 6.4, $\mathbf{t}_1' = [-0.979 \ -0.204]$ after normalization. Consequently $\mathbf{u}_1' = [0.204 \ 0.979]$, for $\mathbf{t}_1 \mathbf{u}_1$ must $= 0$ as OU is at right angles to

TABLE 6.4

Constructs	W		t_0	$l_1 t_1$	t_1	$l_2 t_2$	t_2	
6	24	−5	1	24	1.0	25.05	1.0	etc.
13	−5	12	0	−5	−0.21	−7.52	−0.30	

OT. The two vectors are combined in what may be called a transformation matrix,

$$T = \begin{bmatrix} 0.979 & 0.204 \\ -0.204 & 0.979 \end{bmatrix}$$

and the required values are obtained from $T'WT$ as

$$\begin{bmatrix} 25.50 & -2.19 \\ -2.19 & 10.50 \end{bmatrix}$$

Thus the variation along the axis OT has increased to 25.50, the variation along OU has diminished to 10.50, and the correlation between the two variables is reduced to −0.134.

Briefly, as OT and OU rotate towards OX and OY respectively the variation along OT increases and becomes more independent of the variation along OU. When it reaches OX it is maximized and is completely independent of the variation along OU, which has simultaneously reached OY and is minimized. The two axes are strictly orthogonal, being mathematically independent as well as geometrically at right angles (details of the measurements along them have been given in Table 6.4).

The variation along OY is the residual left after the variation along OX has been extracted. Being minimized, it satisifies the criterion of the method of least squares. When the grid describes variation in three dimensions the residual variation after the extraction of the first component is minimized and confined to a plane in the three-space orthogonal to the major axis. The extraction of the second leaves a minimal amount of residual variation confined to a third axis orthogonal to the first two. Similarly when the dispersion is k-dimensional the extraction of the components one at a time from largest to smallest leaves minimal amounts of variation confined to hyperplanes of one dimension less each time, always orthogonal to the axes of the components extracted.

Thus an incomplete principal component analysis is efficient as far as it goes. The requirements of the method of least squares are satisfied at every stage. If a single scale is needed to represent the observations in a grid approximately and deviations from it have to be neglected, the major axis will give the closest approximation and neglect as little as possible of the observed variation. The total, V, is broken down into the variation along the major axis, l_1, and the residual, $V - l_1$. Again, if two scales are needed the first two components give the closest fit and the residual is reduced by l_2. A complete, exhaustive analysis leaves no residual at all: it accounts for the whole of the observed variation by referring it to the k components.

Principal component analysis is unique in these respects. Any other method of analysis which produces different results cannot satisfy the criterion of least squares. It will leave some amounts of variation remaining in dimensions from which they should have been extracted. Accordingly, the examination of the evidence cannot be systematic and exhaustive. Moreover, it will be obstructed because the reference axes ('factors') supplied in place of those of the principal components will not be strictly orthogonal, and variation along them will be confounded. Such axes are also liable to be unstable under operations related to Hotelling's iterative procedure, e.g. calculating factor scores. Instead of being defined precisely in mathematical terms they are generally interpreted and 'identified' verbally in obscure equivocal language. In short there are definite, substantial disadvantages in all such methods, and when examined carefully the compensating advantages they offer may be found to be nebulous.

7

THE COMPLETE ANALYSIS OF AN INDIVIDUAL GRID: PRELIMINARY PROCESSES AND RESULTS

7.1 Complete analysis

An analysis of a grid may be described as complete if it accounts for every entry exactly. Complete analyses may not often be required. Sometimes a clinician may be willing to examine all his results with open-minded curiosity, acting upon the precept of Plautus, *nihil humani alienum puto*. An experimental psychologist is generally more likely to be guided by some hypothesis which directs his attention to one particular part — possibly only a single measurement. For instance, he might just pick out the standardized distance between self and ideal self, because he expects to find it greater in one group of cases then another. Yet even though a complete analysis were never required a program that can provide one may have many uses, satisfying different needs on different occasions. It is not safe to assume that any part of the evidence in a grid is invariably void of psychological significance.

Some procedures such as analysis of variance and factor analysis, which prescribe computations that can be applied to grids, are intentionally incomplete. They exclude part of the observed variation, refer to it as residual and attribute it to error. Admittedly the data in a grid may be exposed to errors of observation; one may doubt, however, whether the conglomerate can be refined into pure truth and pure error by computations. And besides, psychology is not bound to turn back at the borderline between truth and falsehood. Factitious illness, pseudologia fantastica and the Munchausen syndrome are familiar even to the general practitioner. (Note on *Factitious* illness: A non-existent illnes may produce genuine symptoms: e.g. a girl may suffer from morning sickness although she is not pregnant. The term factitious is applied to such conditions.) Though there may be no grain of truth in a grid, treating it as rubbish may be a mistake. So one may prefer to view the whole of the evidence as potentially worth studying and avoid methods of analysis which preclude parts of it from consideration automatically.

The analysis of a grid, say **G**, into its principal components can be made exhaustive. To complete it in the simplest possible way **G** must first be centred. This converts it into **D**, the array of deviations from the construct means, defined by its characteristic entry as

$$d_{ij} = g_{ij} - g_{i.}$$

Then **D** is resolved into its principal components. The relationship is expressed by the formula

$$\mathbf{D} = \mathbf{C} \mathbf{L}^{\frac{1}{2}} \mathbf{E}'$$

where C is the matrix of construct vectors of the components, L the diagonal matrix of their latent roots and E the matrix of their element vectors. All these terms were defined explicitly in Chapter 6 except E, which was only defined implicitly when the product $C'D$ was calculated. The connection can be stated as

$$C'D = L^{1/2} E'$$

The properties denoted by the terms g_i, C,L and E are essential in that the principal component analysis of a grid is not complete unless they are all specified. Other properties can be derived conveniently during the course of the analysis or after it is completed. The INGRID 72 program, which appears to be becoming the standard one for analysing grids individually, lists a good many derived properties which are commonly found of psychological interest. But of course there is no limit to the number that can be derived in one way or another, and the possibility of finding some further formula for a measurement of psychological interest can never be precluded entirely. In this sense no analysis of a grid can be described as complete.

The program will analyse any number of individual grids one after another in sequence. The elements may be ranked or graded in terms of the constructs. Any scale may be used for grading. For instance, it may be dichotomous, i.e. an all-or-none scale, as favoured by users of Kelly grids; or a seven-point scale, as favoured by users of Osgood 'semantic differential' grids; or a percentage scale, etc. It is convenient to keep to the same scale for all the constructs in a grid and essential to record an evaluation of every element in terms of each of the constructs. For instance, if ranking is used all the elements should be ranked in terms of each of the constructs. (Ranking reduces to grading when ties are allowed; that is to say, grids with tied ranks are analysed as graded.)

The output can be varied to suit the needs of the user. Some parts which are regularly listed may be excluded if he does not need them. For instance, if he is only interested in the correlations between the constructs or the distances between the elements he can exercise the option to exclude the whole of the principal component analysis. Other parts, not regularly listed, can be included if he wishes; thus by using another option he can get a listing of tables of the residual deviations after the extraction of each component for as many components as he likes up to k. There are eight options in all.

A pilot card precedes each grid when it is read in, giving the number of constructs and elements, the rating method used and the options selected, with some other information to identify it. So grids with different numbers, methods and options can be put together in a single batch for the computer. As soon as a grid has been read in it is written out to provide a check for the user.

This chapter formulates the operations performed by the program and illustrates the results it produces before extracting the principal components. Besides centering, the operations include the optional process of normalization and the calculation of the correlations and angular distances between the elements. The distances are derived from the product matrices DD' and $D'D$, one of which is needed for the component analysis.

Some of the operations only apply to graded grids. Their effects will be illustrated here by applying them to the addict's grid (Table 3.1). The other preliminary operations that apply to all grids will only be described formally here.

Their effects will be illustrated by data from a ranked grid, presented and discussed in Chapter 9.

7.2 The means, variation, etc., of the constructs in a graded grid

The output from a graded grid begins with a definition of the grading scale, reproduced from the pilot-card. Then the grid is listed to avoid any uncertainty about the data being analysed. Next comes a list giving the mean for each construct and the total variation about it, expressed both as a sum of squares and as a percentage of the total variation in the grid ($g_{i.}$, V_i and $100 V_i/V$ for $i = 1, \ldots, n$). The total, V, follows, and the preliminary results conclude with the measures of bias and variability which are defined and discussed in the next section.

If there is any construct in the grid that makes no distinctions between the elements, putting them all in the same grade and contributing nothing to V, it is excluded from further analysis; the value of n is reduced by 1 and the constructs are renumbered to run consecutively from 1 up to the reduced value of n. A statement that this has been done appears in the listing.

Table 7.1 lists the results derived from the addict's grid. Construct means usually deviate from the midpoint of the scale to some extent. In this case the deviations are mostly in the direction of the upper limit and some are distinctly greater than others. Constructs 2, 3, 7, 8, 11 and 14 have no entries below the midpoint; the informant has been able to discriminate between the drugs in the extent to which they produce the effect described by each of these constructs, but has not recorded that any of them counteract it. The most marked of the effects, on the average, are to make him feel high and blocked and give him a good buzz; making him imagine things and see or hear people who aren't really there are more specific or not so

TABLE 7.1   Results derived from the addict's grid. Highest grade on construct scale is 1 and lowest grade is 5

Construct	Mean	Variation	As percentage
1	2.44	18.22	11.71
2	1.56	4.22	2.71
3	1.56	4.22	2.71
4	3.22	25.56	16.43
5	2.11	8.89	5.71
6	3.00	24.00	15.43
7	2.22	5.56	3.57
8	2.89	0.89	0.57
9	2.67	8.00	5.14
10	2.11	6.89	4.43
11	1.67	8.00	5.14
12	2.89	24.89	16.00
13	3.67	12.00	7.71
14	2.44	4.22	2.71

Total variation about construct means 155.56

Bias   0.4093
Variability 0.5893

strong and a slight feeling of sickness is induced by cannabis but not by any of the other drugs.

There are remarkable differences in the amount of variation about the means of the constructs. Constructs 4, 6 and 12 have the largest and construct 8 hardly any — showing that the drugs differ most in the extent to which they affect feelings of sleepiness, drunkenness and tension and least in the extent to which they induce a feeling of sickness. It could be argued that construct 8 is practically useless because it makes so little distinction between the drugs. But this may not justify ignoring its implications; the clinician may be just as interested to learn that none of the drugs produces any marked feelings of sickness as that they vary widely in other respects.

7.3 Bias and variability

The next measurements listed in the output for graded grids are those of bias and variability: *bias* records a tendency for responses to accumulate at one end of the grading scale and *variability* a tendency for them to gravitate towards both ends, leaving the intermediate grades relatively empty.

The amount and direction of the bias in the evaluation of the elements in terms of one of the constructs, say i, can be measured by $b_i = g_{i.} - p$, where p is the midpoint of the grading scale. The variation about the mean is V_i. In accumulating these measurements for a grid as a whole allowances must be made for differences in the orientation of the constructs, for the number of constructs in the grid and for the range of the scale. The cumulative measure of bias is accordingly defined as

$$b = \frac{(\Sigma(g_{i.} - p)^2/n)^{\frac{1}{2}}}{q} \quad (i = 1, \ldots, n)$$

and that of variability as

$$y = \frac{(V/n(m-1))^{\frac{1}{2}}}{q}$$

where q is the distance between p and the limits of the scale.

The formula for bias defines how far the centre of the dispersion of the elements deviates from the midpoint of the construct-space; to define the direction of its deviation the n value of b_i would need to be used as a vector.

These measurements characterize the grid as a whole. They are suitable for comparing it with other grids which have the same number of elements, graded on scales with the same range, but they are not suitable for internal comparisons between different parts of one grid. In other words, they are macrocosmic not microcosmic.

They may be used, for instance, to compare informants who have completed similar grids. In this respect they are analogous to the measures of response sets obtainable from attitude scales. Suppose a scale containing m statements expressing various social attitudes is completed by n informants, recording their responses over a range from 'agree strongly' to disagree strongly'. The data obtained can be arrayed as an n by m collective grid, with a row for each informant and a column for each statement. Under these conditions the quantity b_i will measure the strength of informant i's tendency to agree or disagree with every statement, regardless of its content, and the quantity V_i his tendency to use or avoid extreme responses.

The occurrence of such tendencies in responses to attitude scales is well attested (see Berg, 1967), and various indices for measuring them have been used which are approximately the same as b_i and V_i. In the context of trait psychology it seems natural to refer them to personal characteristics, taking bias, say, as a measure of acquiescence and variability as a measure of extremism.

Whether they should be interpreted similarly in the context of gird technique is open to question. Whereas every row in the collective grid provides one measure of each kind for a different subject, every row in the individual grid provides a different measure of each kind for the same subject. If they are consistent throughout the grid the cumulative measurements, b and y, evidently characterize his construct system as a whole and may perhaps be interpreted as indices of personality traits. But the drug addict's grid provides a cautionary example of how far the measures may vary from construct to construct in an individual grid. (If the problem whether the constructs have been used consistently is considered sufficiently important it would be possible to apply tests of homogeneity borrowed from analysis of variance).

The interpretation of the two measurements is complicated further because of the variety of ways in which grid technique can be used. It is unsafe to assume that variables which have the same formal definition must have the same psychological content; and collateral information is desirable to judge whether any interpretation in general terms applies to the measurements obtained from a particular grid. Experimental evidence on possible implications of bias and variability has been assembled and discussed by Chetwynd (1974 and also Chapter 13).

In some contexts the measurements may be ignored as nuisance parameters which merely need to be excluded before the important contents of a grid can be located.

7.4 The array of deviations

The array of deviations from the construct means, **D**, to which all subsequent analyses refer, follows next for all grids whether graded or ranked. The entries after normalization are given for a graded grid when normalization has been requested.

In this case the characteristic entry is

$$d_{ij} = (g_{ij} - g_{i.})V_i^{-\frac{1}{2}}$$

In a graded grid which has not been normalized, it is simply

$$d_{ij} = g_{ij} - g_{i.}$$

When the elements are ranked, every row of **G** contains a permutation of the first m natural numbers. So all the constructs have the same mean and variation, namely

$$g_{i.} = \frac{m+1}{2}$$

and

$$V_i = \frac{m^3 - m}{12}$$

Normalization would thus be futile and nothing could be learned from these terms about differences between the constructs. INGRID omits them and derives **D** from the simpler formula above.

The entries in a row show how different elements are evaluated in terms of one construct, while the entries in a column show how one element is evaluated in terms of different constructs. Information of the first kind may be gathered as easily from the original grid **G** as from **D**, but it is easier to gather information of the second kind from **D**. The entries in one column of a graded grid may not be directly comparable because they refer to constructs which may differ in their means and variation.

7.5 The effects of normalization

Normalization, in assigning an equal weight to every construct, reduces the extent of variation along the axes of constructs where it was relatively large originally and expands it where it was small. Whether such alterations in scale are justifiable is a question that cannot be answered in universal terms.

> *Pro.* Psychologists, only too familar with arbitrary incommensurate scales of measurement, favour standardization (which is equivalent to normalization) as a general rule.
>
> *Contra.* One should not tamper with the evidence. Grid technique offers the informant a common scale for all the constructs, and if he reports wider variation on some than others presumably they are the ones he finds more effective for discriminating between the elements.
>
> *Pro.* Wider variation on some constructs may only reflect extreme responding; they may be ones where he finds it difficult to make fine distinctions and just places the elements at one end of the scale or the other.

The debate is inconclusive.

The form of the dispersion of the elements in the construct-space is bound to be altered when variation is expanded along some of its reference axes and contracted along others. So the decision to normalize or not must affect all the results obtained from the analysis of **D** with the exception of the table of correlations and angular distances between the constructs, for which normalization is obligatory.

The person who submits a grid for analysis by INGRID is left to decide whether it should be normalized or not. It may not matter much which way the decision goes if the grading scale has been applied consistently and the constructs do not differ greatly in their variation. In case of doubt the S^2_{max}/S^2_{min} ratio (Pearson and Hartley, 1958) could be used to test whether the differences are negligible. If they are substantial some explanation is needed for them, and it may not be evident from an examination of the grid itself. The person who submits it is the one most likely to have access to further information. He may well be someone with clinical experience, who knows the informant personally, has conducted the interview and recorded the grid, and who will finally be responsible for interpreting the results. If he does not exercise his option, the customary one, to normalize, is taken.

There may be good reasons for the unusually large differences in the variation of the constructs in the addict's grid: some constructs may reveal wider differences

TABLE 7.2   Deviations from the construct means in the addict's grid, where 1 is before and
11 after normalization

Effects	Alcohol		Barbiturates		Cannabis		Cocaine	
	1	11	1	11	1	11	1	11
1. Talk more	0.44	0.10	0.44	0.10	−1.56	−0.36	1.44	0.34
2. Feel high	−1.44	−0.70	0.56	0.27	−0.44	−0.22	0.56	0.27
3. Blocked	−1.44	−0.70	0.56	0.27	−0.44	−0.22	0.56	0.27
4. Sleepy	0.22	0.04	2.22	0.44	1.22	0.24	−1.78	−0.35
5. Warm inside	0.11	0.04	1.11	0.37	0.11	0.04	1.11	0.37
6. Drunk	1.00	0.20	2.00	0.41	1.00	0.20	−2.00	−0.41
7. Imagine things	−0.78	−0.33	−0.78	−0.33	1.22	0.51	−0.78	−0.33
8. Sick	−0.11	−0.12	−0.11	−0.12	0.89	0.94	−0.11	−0.12
9. Act bluntly	−0.33	−0.12	1.67	0.59	−0.13	−0.47	−0.33	−0.12
10. Enjoy things	−0.89	−0.34	0.11	0.04	0.11	0.04	0.11	0.04
11. A good buzz	−1.33	−0.47	0.67	0.24	0.67	0.24	0.67	0.24
12. Tense	−0.11	−0.02	−2.11	−0.42	0.89	0.18	1.89	0.38
13. Sexy	1.67	0.48	−1.33	−0.38	1.67	0.48	−0.33	−0.10
14. Hallucinated	−0.56	−0.27	−0.56	−0.27	−0.56	−0.27	0.44	0.22

between the drugs and some may have more easily intelligible bipolar contrasts than
others. The decision not to normalize might be justified in this case.

Table 7.2 illustrates what the effects would be on the array of deviations. The
first four columns of the array before and after normalization are tabulated. As the
differences in variation are very wide the effects on the relative size of the entries
for an element must be exceptionally great. One may notice, for instance, that
while the entry for alcohol on construct 6 is reduced from 1.00 to 0.20 its entry on
construct 7 changes from −0.78 to −0.33.

But on the whole it may seem surprising that the effects are no greater than they
are. The rank order of the entries for an element is only changed at a few points. It
is clear that the effects are bound to be limited. Only the relative magnitude of the
entries can be altered; their signs must remain the same since they are determined
by centering. Therefore an element must remain in the same sector of the
construct-space — a sector which is $(\frac{1}{2})^n$th part of the total. So one may expect
that the effects of normalization will become slighter as the number of constructs
increases.

7.6 Corresponding results for a ranked grid

The corresponding part of the output for a ranked grid is limited to the tables of **G**
and **D**. Everything else is irrelevant. **D** is defined as

$$d_{ij} = g_{..} - g_{ij}$$

where $g_{..}$ is the mean of the first m natural numbers, $(m + 1)/2$. By this definition
positive values of d_{ij} are assigned to elements ranked high and negative values to
ones ranked low.

All the results that follow are put in the same form for all grids.

7.7 The correlations and angular distances between the constructs

When **D** has been obtained from a graded grid without normalization the matrix of correlations between the constructs needs to be calculated from

$$\mathbf{R} = \mathbf{V}^{-\frac{1}{2}} \mathbf{D}\mathbf{D}' \mathbf{V}^{-\frac{1}{2}}$$

When the elements have been ranked all the entries in **V** are the same, say v, and the calculation can be simplified to

$$\mathbf{R} = v^{-1} \mathbf{D}\mathbf{D}'$$

When the constructs have been normalized they all have a variance of 1, so the equation simplifies further to

$$\mathbf{R} = \mathbf{D}\mathbf{D}'$$

The complete matrix **R** is a symmetrical square matrix with units in its leading diagonal. INGRID 72 only calculates the entries above the leading diagonal; it lists the correlation between each pair of constructs, r, followed by the angular distance between them, α. The relationship is $\alpha = \cos^{-1} r$. For an example of a table of correlations see Table 9.8 (angular distances have not been included there).

Often there is much to be learned about the informant's construct system from examining the correlations between the constructs. Indeed, they are sometimes treated as the only results from a grid that need detailed attention. The angular distances between the constructs, which are printed alongside the correlations between them, describe their dispersion in the element-space (E-space). An explanation of this may now be needed.

7.8 The two distributions specified by the data

As already mentioned in Chapter 5, grids may be read by row or by column. For example, reading along row 2 of the addict's grid and comparing the entries in the first two columns we see that cannabis is more effective than alcohol in making him feel high; reading down column 2 and comparing the entries in rows 2 and 7 we see that while cannabis makes him feel high it does not make him imagine things. Each way of reading the arrays gives the specifications for a multivariate dispersion. The entries in any column form a vector of coordinates giving the location of an element in a space with an axis for every construct — say the C-space for short. Likewise the entries in any row form a vector locating a construct in a space where there is an axis for every element, say the E-space.

The properties of a dispersion of elements in a C-space have already been considered in Chapter 6. It is a scatter, approximately elliptical in form, of m points in an n-dimensional space. When **G** is centered and converted to **D**, the point of equilibrium of the dispersion, the multivariate mean $g_{i.}$, is transferred to the origin of the C-space. Otherwise its form is unchanged.

The dispersion of the constructs in the E-space, which is to be considered now, has quite a different form. It consists of n points. They are scattered in an m-dimensional space, but are excluded from one dimension of it by centering. This implies that

$$\mathbf{D1} = \mathbf{0} \quad \text{(where } \mathbf{0} \text{ is a column vector of } n \text{ zeros)}$$

The column vector 1 of m units thus defines a dimension in the E-space where there is no variation. Although the two dispersions are very different in appearance they are connected becaues they both describe the same array of data.

No matter what method has been used for evaluating the elements, the variance of every construct is implicitly reduced to 1 when the correlations between them are calculated. Consequently, they are all placed at an equal distance from a common origin in the E-space and differ only by scattering away from it in different directions. They lie on the surface of a hypersphere and the difference between any two of them can be expressed as an angular or circumferential distance — the angle they subtend at the centre.

An angle of $0°$ corresponds with a correlation of $+1.0$. It implies that the constructs are located at the same point on the hypersphere. An angle of $90°$ corresponds with a correlation 0.0. It implies that the constructs are independent of one another. An angle of $180°$ corresponds with a correlation of -1.0. The two constructs are located diametrically opposite to one another; one provides the same scale of measurement as the other, but in reverse.

In some contexts it is an advantage to consider the angular distances between constructs rather than their correlations: the average of a set of angles is itself an angle whereas the average of a set of correlations is not itself a correlation. So the angle corresponding to each correlation is printed out alongside it. Such measurements can be used for comparing girds. For instance, it would be possible to compare the average angular distances between the constructs 'Like me as I am' and 'Like I would like to be' in the grids of informants from different groups without using the same elements for all the grids or keeping the other constructs the same.

7.9 Corresponding properties of the elements

The contents of **D** are reexamined next to extract information about the elements. The totals of the entries for every element are calculated and listed and so are the sums of their squares. The results from the grid analysed in Chapter 9 are given in Table 9.3.

In general, the sum of squares for element j will be denoted S_j. It is convenient, for reasons which will become apparent later, to write the complete set of sums of squares for the m elements as a diagonal array S. Their total, being the sum of the squares of all the entries in **D**, is equal to the total variation about the construct means, i.e $S = V$. If normalization has been applied, $S = V = n$; and if ranking has been used, $S = V = n(m^3 - m)/12$. In either of these cases it is listed as the total variation about the construct means, and the total per construct, which, correspondingly, will either be 1 or $(m^3 - m)/12$, is also given. If S comes from a graded grid that has not been normalized it is equal to the value of V listed after the mean and variation of the constructs (see Table 7.1), so it is not listed again.

The analysis of the total variation, thus defined, is one of the main concerns of the program, and the sum of squares for each element is listed as a percentage of it. The importance of any element in the construct system is indicated by the proportional size of its sum of squares. The quantity $S_j^{1/2}$ measures the distance of element j from the centre of the dispersion in the C-space. If it is small the informant must have rated the element neither high nor low but near the mean on all the consttucts. This would suggest that his attitude towards it is indifferent. Conversely, a salient element, one with a relatively large sum of squares, would appear to be an important one in his construct system, whether his attitude towards it is consistently favourable or consistently unfavourable, or favourable in some respects and unfavourable in others.

It may be well worthwhile to see which are the salient elements when attempting to interpret the contents of a grid. When a few account for a large proportion of the total variation it is to be expected that their location in the C-space will determine the orientation of the major axes of the dispersion there. In this case the psychological contents of the components may be discerned more readily by referring them to the elements than to the constructs which contribute to them.

The totals for the elements are not necessarily as useful a guide to the more important contents of a grid as their sums of squares. One construct may approximate to the contrast of another, and correlate negatively with it. Then although the differences between the elements may show up clearly on each construct separately, their totals on the two may be much the same. Their totals on all the constructs will locate them on the axis in the construct-space defined by a row vector of n units, 1, for they can be specified by the expression 1D. The direction of this axis is arbitrary, depending on which poles or the constructs are treated as the emergent ones as well as on the choice of constructs. The totals are listed nevertheless because they may be useful for some computing purposes.

Premultiplying D by its transpose yields an m by m matrix, $P = D'D$, which is put to various uses. It contains S as its leading diagonal; the typical off-diagonal entry is denoted P_{ij} here. An option is available for obtaining an abbreviated listing of P which just gives the values of P_{ij} for i, j $(i < j)$.

The distance between two elements, say J and K, measured on the scale which was adopted for recording D, is properly defined as

$$(\Sigma(d_{ij} - d_{ik})^2)^{\frac{1}{2}} \qquad (i = 1, \ldots, n)$$

but can also be calculated conveniently from P as

$$(S_j + S_k - 2P_{jk})^{\frac{1}{2}}$$

It can be expressed on a standard scale if it is compared with the expected distance between a pair of elements taken at random from the same grid. The expected value for S_j and S_k is the average of the entries in S, namely S/m. Similarly the expected value for P_{ij} is the average of all the off-diagonal entries in P, and this must be $-S/m(m-1)$ since the sum of the entries in every row and column of P is 0 (see Slater, 1951, for a proof). Inserting these values into the equation for the distance between J and K we obtain the quantity

$$(2S/(m-1))^{\frac{1}{2}}$$

as the expected distance between two elements taken at random. It is given as the 'unit of expected distance' in the output from the program.

The distance between each pair of elements proportionate to this unit is listed next; thus distances recorded as over 1 are greater than expected and those under 1 less. These standardized distances may be used in comparisons between grids even if they do not all refer to the same elements or constructs. For example, if 'Me as I am' and 'Me as I would like to be' are included as elements in grids with a common focus of interest obtained from different groups of informants, it would be possible to see whether the distances are wider in one group than another. The grids would not need to be matched in other respects.

There is an option to conclude the analysis at the point. If it is not taken the program goes on to a principal component analysis.

8

THE COMPLETE ANALYSIS OF AN
INDIVIDUAL GRID: FINAL PROCESSES
AND THE SEQUENCE OF RESULTS

8.1 Introduction

Some relatively unfamiliar extensions of principal component analysis are used by INGRID 72 in accounting completely for the array of deviations **D**. This chapter starts by explaining the methods adopted and the reasons for them, and concludes by describing the sequence in which the results are listed in the output from the program. Illustrative results are presented and discussed in Chapter 9.

8.2 Principal component analysis applied to a grid

The general nature of principal component analysis was outlined in Chapter 6, taking an array of data recording a dispersion in two dimensions as an example. When the dispersion extends into more dimensions the mathematical problem of computing its latent roots and vectors becomes rapidly more complicated. Wilkinson (1965) has written a comprehensive account of its treatment. In INGRID 72 the computations are carried out by an updated version of subroutine HOW, which was written by David W. Mantula at the Berkeley Computing Centre, University of California, in 1962, using the methods developed by Householder, Ortega and Wilkinson. By now it has been used for analysing many thousands of grids and has proved completely satisfactory.

The matrix to which the subroutine is applied is the covariance matrix between the elements **P**. It is condensed into a tri-diagonal matrix with the same latent roots. Thence its k roots are all obtained simultaneously by direct solution of the determinantal equation. Finally, the element vectors are derived and stored as an m by k array, **E**.

The subroutine is applied to **P**, not **W**, as it is usually the smaller matrix since grids generally contain more constructs than elements. The results would be the same if it were applied to **W**, i.e. to $\mathbf{DD}'$, instead of **P**, which is $\mathbf{D}'\mathbf{D}$, because every non-zero latent root of **P** is also a root of **W**. Thus component i, derived from **P**, with its latent root l_i and element vector $\mathbf{e}_i$, will also be derived from **W** as component i; its latent root will be l_i and its construct vector will be proportional to $\mathbf{De}_i$. Its characteristic equation with reference to **P** is

$$\mathbf{D}'\mathbf{De}_i = l_i\mathbf{e}_i$$

Premultiplying both sides by D gives

$$DD'De_i = Dl_ie_i$$

which may also be written as

$$WDe_i = l_iDe_i$$

(l_i being transferable since it is a scalar). And this is the characteristic equation for component i with reference to W, otherwise written as

$$Wc_i = l_ic_i$$

So De_i must be proportional to c_i, and will be identical when both are normalized. The factor needed for normalization is $l_i^{-½}$. (Note that the vectors denoted e_i and c_i appear as the entries in column i of E and C respectively, and might therefore be indicated more precisely as $e_{.i}$ and $c_{.i}$.)

As c_i can be obtained from De_i for all non-zero values of l_i there is no need to analyse W as well as P. The complete set of construct vectors can be obtained from

$$C = DEL^{-½}.$$

Alternatively, if W is analysed there is no need to analyse P. The construct vectors of the components will be obtained first and their element vectors can be derived from the formula

$$E = D'CL^{-½}.$$

Since every component has a construct vector as well as an element vector the entire component-space must be contained within the C-space as well as the E-space. It describes the connection between the two: the orientation of its axes in the C-space is defined by the construct vectors of the components and in the E-space by their element vectors. The variation along the axis of a component in the C-space is also the variation along its axis in the E-space, namely its latent root.

There is an entry for every element in the element vector of every component. And as their total must be zero some of the entries must be positive and some negative. That is to say, components are bipolar just as constructs are, and include all the elements within their range of convenience. Likewise there is an entry for every construct in the construct vector of every component (the entries will differ in magnitude and probably also in sign though their total need not be zero). Thus every component includes all the constructs as well as all the elements within its range of convenience.

As the components with non-zero latent roots account for all the variation recorded in D, there can be no variation outside the component-space either in the C-space or the E-space. Therefore grids contain no evidence of the occurrence of specific factors. Any analysis or test of significance which assumes their occurrence is inappropriate by definition to all data in the form of grids.

The equations

$$De_i = l^{½}c_i \quad \text{and} \quad D'c_i = l^{½}e_i$$

which define the relationship between e_i and c_i can be premultiplied alternatively by D and its transpose any number of times without producing any change in the normalized vectors c_i and e_i (unless by errors of computation). The characteristic equations imply this. But it is only the vectors of the principal components that have this stability. If we take any other normalized vector of m terms, say x, and premultiply it by D, we will obtain a vector of n terms, say y, which can be premultiplied by D' to obtain another vector of m terms, say z. When z is normalized it will prove different from x. What has happened is that Hotelling's iterative process for finding the major axis of a dispersion has been brought into operation, and z will approximate closer to the major axis than x. Moreover, the variation along y will not be the same as along x or z, but intermediate in amount. This instability complicates the use of any vectors other than those of the principal components for relating measurements in the C-space to measurements in the E-space (e.g. calculating factor scores from factor loadings of measurements from any set of rotated axes).

The elements can be located in the component-space by their loadings, which may be defined collectively as $D'C$ or $EL^{1/2}$. The loadings of element j,

$$l_1^{1/2}e_{j1}, \ l_2^{1/2}e_{j2}, \ \ldots, l_k^{1/2}e_{jk}$$

which are the entries in row j of this m by k product matrix, form the vector giving the location of element j. This reference system is easiest to understand when the component space is regarded as contained in the construct space — a space where the constructs function as axes and do not appear as points (see Figure 9.1).

The constructs, however, can also be located in the component-space. Their loadings may be defined collectively as DE or $CL^{1/2}$, which is an n by k matrix. The vector giving the location of construct i is the set of entries in row i,

$$l_1^{1/2}c_{i1}, \ l_2^{1/2}c_{i2} \ \ldots, \ l_k^{1/2}c_{ik}$$

This reference system is easiest to understand when the component-space is regarded as contained in the element-space, where the elements function as axes and do not appear as points.

Since elements and constructs can both be located in the component-space, it might seem reasonable to draw maps of sections of the space by taking any pair of components as the reference axes and proceeding to mark both the elements and the constructs as points on the surface, using their loadings as the coordinates for locating them. Such maps would give the impression that the relationship between an element and a construct can be expressed as a linear distance. Indeed it would be mathematically possible to use the results from the analysis to calculate the linear distance between an element and a construct. (The method employed for calculating linear distances between pairs of elements, see Section 7.9 could be applied to the differences between their loadings on each of the components.)

But linear distances are not suitable for expressing relationships between constructs and elements because their proportionate distances from the centre of the component-space depend on their numbers. Consider starting with the smallest size of grid that can have k components, namely one with $n = k$ and $m = k + 1$. Now let us introduce more constructs with about the same amount of variation as the previous ones. They will occupy more points in the component-space at about the same distance from its centre, so the effect will be to increase the density of their dispersion there. The total variation recorded in the grid will also increase, and as the number of elements is unaltered the total per element will increase; so the corresponding effect on the dispersion of the elements will be to expand it. Conversely, if more elements are introduced while the constructs remain the same, the dispersion of the constructs will expand while that of the elements condenses. (None of these operations will affect the dimensionality of the component-space.)

Though the linear distance between two points must change when one moves further away from an origin while the other remains stationary, the angle they subtend at the centre need not change. So it seems advisable to express relationships between elements and constructs either as angular distances or measurements which are simple functions of angular distance. As there is a point for each of the elements and constructs in the component-space, the relationships between all of them can be quantified in this way.

Actually the most convenient procedure is to use the formula for correlation to calculate cosines and to derive the angular distances from them. The formulae stated with reference to the axes of the components would be

$$R(c, c) = V^{-\frac{1}{2}}CLC'V^{-\frac{1}{2}}$$

for the cosines or correlations between the constructs and

$$R(e, e) = S^{-\frac{1}{2}}ELE'S^{-\frac{1}{2}}$$

for the cosines between the elements. The cosines between the constructs and the elements can be calculated similarly as

$$R(c, e) = V^{-\frac{1}{2}}CLE'S^{-\frac{1}{2}}$$

The calculations required by the first two formulae are unnecessarily laborious. The correlations between the constructs can be obtained in the usual way (see Section 7.7) without referring to the components. The results are exactly the same because

$$C'LC' = DD'$$

The cosines between the elements can also be obtained more conveniently from

$$R(e, e) = S^{-\frac{1}{2}}D'DS^{-\frac{1}{2}}$$

since

$$ELE' = D'D$$

But there does not appear to be any way of obtaining the cosines between the constructs and the elements without referring to the axes of the components.

$R(c, c)$ and $R(e, e)$ are symmetrical square matrices which only need to be specified by the entries above their leading diagonals, while $R(c, e)$ is an n by m matrix of which every entry needs to be listed. The three can be put together in a simple triangular array looking as if it were the set of entries above the leading diagonal of a symmetrical $(n + m)$ square matrix, (see Table 9.8).

Which are preferable, degrees or cosines, as terms for describing the angular distance between the functions? The question cannot be answered decisively. Correlation coefficients are by far the most familiar terms for describing relationships between variables in psychological contexts, so cosines, which are mathematically equivalent to them, are likely to be found the most readily acceptable.

Unfortunately their use may lead to some misunderstandings. Correlations have other properties besides being cosines. They are used in statistics to measure the regression of one standardized or normalized variable on another, and their standard errors may be calculated from the formula

$$\text{s.e.} = \left(\frac{1 - r^2}{n - 2}\right)^{\frac{1}{2}}$$

where n is the number of cases observed. But while we have found that functions of both kinds in grids can be treated as variables we have also found that the total variation recorded for each is confined to k independent components. This is one reason for hesitating to assess the significance of the cosine between any pair, even when both are constructs and the number of elements is entered as n. The application of probability theory to the contents of a grid also involves more fundamental problems (see Chapter 10).

Such problems do not arise so long as cosines are used simply for the purpose of measuring angular distances between pairs of axes, and this is the intention with which they are used in Table 9.8. But degrees are equally suitable for this purpose, and if they were brought into general use and became as familiar terms as correlations they might be found to have some advantages. Their definition is more elementary and easy to understand, they have acquired no misleading attributes and they have many advantages for use in further operations, e.g. comparing measurements from different sources.

Fisher's z-transformation has no advantages in this context. Apart from being a statistic for testing significance and not just a measurement, which is all that is required, it has the disadvantage of receding to infinity as r reaches 1.0, and as r can do so in grids, infinite values of z would actually occur, making further computations, e.g. calculations of averages, impossible.

8.3 The results in order of presentation: the latent roots

The results from the principal components analysis made by INGRID 72 begin with the statement, 'The component-space is limited to k dimensions', where k is the number of non-zero latent roots found. Then the roots are listed from the largest to smallest as observed quantities and as percentages of the total variation.

Usually just a few of the latent roots — sometimes one, sometimes two, seldom more than three — account for a very large proportion of the total. When this is so, much of the information concerning the relationships of the constructs and the elements with one another can be shown by mapping their dispersion on the axes of the major components. Some of the possibilities are explained and exemplified in the next chapter.

If the option for the Bartlett test is taken, its results are listed next.

8.4 The Bartlett test

Bartlett's test (1950, 1951a, 1951b) is so highly generalized that it can be applied to grids although it was not developed for such data. It referred originally to experiments with R-technique, where a relatively small battery of tests is given to a relatively large number of subjects, who are supposed to constitute a representative sample of some population. The experiment is not concerned with the subjects personally; its object is to estimate the relationships between the tests in the population and to account for their correlations approximately in accordance with the assumption that they measure a smaller number of latent variables to some extent in common.

Relatively large amounts of variation are expected to occur in a few dimensions, leaving a residual amount dispersed at random in the rest of the component-space. The dimensions where the variation is greatest are to be identified one at a time, e.g. by Hotelling's method, and a value of chi-squared can be calculated at any stage to test whether the assumption that the residual variation is randomly distributed has an acceptable degree of probability. For this purpose it is sufficient to know the sum of the latent roots, $l_1 + l_2 + \ldots + l_k$ and their product, $l_1 \times l_2 \times \ldots \times l_k$. The former is given by V, and the latter by $|V|$, provided V is non-singular, as it is almost bound to be in an experiment of the kind envisaged.

To extend the test of grids, where singularities frequently occur and $|V|$ may therefore be zero, all the non-zero roots need to be obtained and their product calculated. The occurrence of any zero roots can then be offset by reducing the degrees of freedom for chi-squared. Thus INGRID, which derives all the latent roots from HOW, can apply the Bartlett test to any grid.

As adapted to the program, the test proceeds from the smaller roots to the larger, first comparing the least-but-one with the very least, then the least-but-two with the two least, and so on, finally comparing l_2 with all the smaller roots combined. Being a test for residual chi-squared it does not extend to l_1. The successive values of chi-squared are listed and the output concludes either by giving the number of latent roots found significant or by printing the statement, 'Negative result from test', if the residual chi-squared after the extraction of the first component is not significant.

Though this procedure provides a solution to the problem of applying the test to grids universally, the appropriateness of the assumptions is open to question when the test is employed in contexts other than those for which it was originally intended. Variation may not only be extended in some directions by experimental conditions but also restricted, intentionally or accidentally, in others. For example, suppose that n people at a mixed gathering are asked to rate an assortment of m cheeses according to personal preference (and suppose $n > m$). If the grid of their responses is normalized by row, as is reasonable to give every person's opinions an equal weight, or even if it is merely centered by row, the dimensions of D will be limited to $m - 1$, and the principal component analysis is bound to reveal one zero latent root.

Now suppose, further, that two very similar bland cheeses have been included in the assortment. Variation along the axis 'mild/strong' will of course be increased, but the experiment will also have provided an extra dimension in the element-space where variation is free to occur but very little is likely to be found. Moreover, the variation in this minor dimension may not be directly concerned with the distinction between the two bland cheeses if this happens to converge with some observable difference between other sorts. The product matrix will then have at least one component with an exceptionally small latent root, and its content may not be easy to interpret.

Thus there are some occasions when the results from the test may appear paradoxical. Whatever the explanation may be, the fact is that extremely enigmatic results — neither readily acceptable nor easily explained away — have occasionally been obtained from the Bartlett test during the course of its use with grids. Perhaps a fresh attempt is needed to consider how the significance and reliability of a grid is to be judged. The subject is discussed in Chapter 10. In the meanwhile, the Bartlett test is still retained in INGRID 72.

8.5 The analysis of the observed variation in terms of the components

The program provides three analyses of the total variation recorded in D: by construct, by element and by component.

The analysis by construct was presented in Section 7.2 which discusses the contribution of each construct to the total

$$V = V_1 + V_2 + \ldots + V_n$$

It is only of interest when the grid is graded and not normalized. Otherwise it is trivial as every construct is bound to have the same variation.

The analysis by element was described in Section 7.9 which discusses the relative salience of the elements indicated by the contributions of their sums of squares to the total

$$S = S_1 + S_2 + \ldots + S_m$$

The third way of analysing the total is as the sum of the latent roots of the components

$$V = S = l_1 + l_2 + \ldots + l_k$$

as mentioned in Section 8.2 above.

Going into further detail, the variation of each construct can be analysed by element and by component, and the sum of squares for each element can be analysed by construct and by component. The analysis of the variation of construct i by element is defined by

$$V_i = d_{i1}^2 + d_{i2}^2 + \ldots + d_{im}^2$$

So the relative importance of the elements on the scale of construct i can be seen by examining the entries in row i of D. Similarly the analysis of the sum of squares for element j by construct is defined by

$$S_j = d_{1j}^2 + d_{2j}^2 + \ldots + d_{nj}^2$$

So the characteristics of element j delineated by the constructs can be seen by examining the entries in column j of D. The listing of D (see Section 7.1) is sufficient for both purposes. The analyses of V_i and S_j by component remain to be considered.

Finally, the evaluation of each element in terms of each construct can be analysed by component. Such an analysis, i.e. the component analysis of each entry in D, is an essential part of the complete analysis of a grid, so it should be understood even it is seldom applied in practice.

We may begin by redefining the array of deviations from the construct means, so far denoted D, as $D(0)$, to indicate that it is the array before any components have been extracted. Then the arrays of residuals after 1 up to k components have been extracted can be denoted correspondingly as $D(1), \ldots D(k)$. The last will consists, of course, entirely of zeros.

Let $D(f)$ and $D(g)$ by two successive arrays in the series $D(0)$ to $D(k)$. Then $D(g)$ can be defined as

$$D(g) = D(f) - c_{.g} l_g^{1/2} e_{.g}$$

The relation between their characteristic entries, $d_{ij}(f)$ and $d_{ij}(g)$ can also be defined as

$$d_{ij}(g) = d_{ij}(f) - c_{ig} l_g^{1/2} e_{jg}$$

The terms in the series $d_{ij}(0)$ to $d_{ij}(k)$ are simple deviations, and may be positive or negative. And though the series must terminate with $d_{ij}(k) = 0$ there may be later terms in it, such as $d_{ij}(g)$, with a larger absolute magnitude than some of the earlier terms such as $d_{ij}(f)$. That is to say, it is possible for the evaluation of a certain element on a certain construct to appear more exceptional after some components have been extracted than before.

When D is redefined as $D(0)$ the diagonal matrices V and S derived from DD' and $D'D$ may be redefined consistently as $V(0)$ and $S(0)$. The corresponding arrays derived from $D(1), \ldots, D(k)$ can then be denoted $V(1), \ldots, V(k)$ and $S(1), \ldots, S(k)$ respectively. It is unnecessary to calculate them by matrix multiplication; the entry for construct i in $V(g)$ can be found from

$$V_i(g) = V_i(f) - l_g c_{ig}^2$$

and the characteristic entry for element j in $S(g)$ from

$$S_j(g) = S_j(f) - l_g e_{jg}^2$$

The analysis of the variation of construct i by component can therefore be defined as

$$V_i(0) = l_1 c_{i1}^2 + l_2 c_{i2}^2 + \ldots + l_k c_{ik}^2$$

and the analysis of the sum of squares for elements j by component can be defined as

$$S_j(0) = l_1 e_{j1}^2 + l_2 e_{j2}^2 + \ldots + l_k e_{jk}^2$$

Since l_g is necessarily positive, being a measurement of variation, there can be no negative quantities in $l_g c_{.g}^2$ for any value of g from 1 to k. So the terms in both the series for $V_i(0)$ and $S_j(0)$ diminish progressively to zero. But the reductions are irregular. That is to say, some of the major components may account for much less of the variation of a particular construct or of the sum of squares for a particular element than one of the minor components. Phenomena of this kind should not be overlooked. They are discoverable in almost every grid and are relevant to its psychological interpretation.

The analyses of the components by construct and by element also involve the quantities typified by $l_g c_{ig}^2$ and $l_g e_{jg}^2$, which are the squares of the loadings of the constructs and the elements. The proportionate contributions of the constructs to the total variation or latent root of component g are indicated by

$$l_g c_{1g}^2 + l_g c_{2g}^2 + \ldots + l_g c_{ng}^2 = l_g$$

and the proportionate contributions of the elements by

$$l_g e_{1g}^2 + l_g e_{2g}^2 + \ldots + l_g e_{mg}^2 = l_g$$

The essential properties of components are their mathematical ones, namely their optimal properties as reference axes for two-way numerical distributions. These are formal and universal. The components derived from a grid are also bound to have psychological properties because the whole of the variation a grid records is psychological in origin. These properties are material and particular. They depend on the choice of constructs and elements and on the state of mind of the informant at the time of the interview, including, one might add, what he expects the interviewer expects of him and much else besides.

They may quite conceivably be similar to the properties of a component in some other grid. But the best way to infer, interpret or identify its pshycological properties (if that task is undertaken) is not often likely to be by analogy with another component from another grid. It is generally advisable to examine the constructs and the elements which lie at the opposite poles of its axis, bearing in mind that it is necessarily bipolar, and then go on to see how the remaining constructs and elements fit in between. The loadings, which give the locations of the constructs and elements on the axis, are the terms to examine for this purpose. Sometimes the distribution of the constructs may be found more illuminating, sometimes that of the elements. An adequate interpretation should always extend to both. Any additional information available about the informant and his psychological predicament should also be taken into consideration.

The relationships between constructs, elements and conponents can also be illustrated by diagrams (see Chapter 9). The loadings, not their squares, are the measurements needed for making them. On this account, and since the square of a number is easier to gauge approximately than its square root, the loadings are the quantities listed in the specifications of a component.

8.6 The specification of the components

The full specifications for a component, say g, listed after the table of latent roots (or the results of the Bartlett test if that has been requested) are

the vector	$\mathbf{e}_{.g}$
loadings	$l_g^{1/2} \mathbf{e}_{.g}$
and residual sums of squares	$\mathbf{S}(g)$

for the elements $j = 1, \ldots, m$, and

the vector	$\mathbf{c}_{.g}$
loadings	$l_g^{1/2} \mathbf{c}_{.g}$
and residual variation	$\mathbf{V}(g)$

for the constructs $i = 1, \ldots, n$. The arrays $\mathbf{S}(g)$ and $\mathbf{V}(g)$, mathematically defined as diagonal matrices, are listed as columns.

The number of components to be specified is optional. The option provides an upper and a lower limit in case the Bartlett test is required. If the number found significant by the test is less than the lower limit, the specifications will still be listed for the components up to it; if the number found is more than the upper limit. the specifications will be terminated when the limit is reached; if the number found lies between the two limits, that will be the number of components specified. If the Bartlett test is not required the limits set as upper and lower should be the same. When no special options are expressed the present practice is to set both limits to 3 and omit the test.

8.7 The polar coordinates

If specifications are obtained for three or more components the lists for the first three are followed by lists of polar coordinates for the constructs and elements, calculated from their loadings on those components. The coordinates can be used for plotting points for the functions (constructs and elements) on the surface of a sphere. This device will indicate the angular distances between them as accurately as three-dimensional space permits. Since the first three components often account for as much as 90 per cent of the total variation in grids of the customary size, a high level of accuracy can often be attained.

The headings used are H, V and R — for horizontal, vertical and radial. The H and V coordinates are the ones used for locating the points. The convention adopted is that the axis of the first component runs from front to back ($0°$, $0°$ to $\pm 180°$, $0°$), that of the second from east to west ($+90°$, 0 to $-90°$, $0°$) and that of the third from north to south ($0°$, $+90°$ to $0°$, $-90°$). Starting from the point on the equator selected as the origin, with $H = 0°$ and $V = 0°$, positive values of H are reached by moving to the right, i.e. eastwards around the globe for the given number of degrees, and negative values by moving westwards. Positive values of V are reached by moving upwards, i.e. northwards, negative southwards. When the point indicated by the coordinates is reached, it can be marked by an adhesive label bearing some identification of the function to which it refers.

The radial measurement R has some interesting properties only indirectly concerned with mapping. It defines the multiple correlation between the function and the three major components; its square measures how much of the total variation or sum of squares of the function they account for. R also indicates how far the vector of the function projects onto the subspace (or hypersurface) described within the entire component-space by the three major components. If it is not greater than 0.7 the function does not project as far there as onto the subspace of the minor components, so its relationship to them should be worth examining. When there are too many functions to be plotted conveniently on the surface of the globe, those with the smallest values of R should be the first omitted.

Since the point defined by H and V does not give the location of the function but only the orientation of its axis in the component-space, a point diametrically opposite would serve to define it equally well. To obtain the coordinates for the opposite point, subtract $180°$ from H when it is positive or add $180°$ when it is negative, and simply change the sign of V.

When the function is a construct the use of two points to indicate its axis is easy to understand: the coordinates listed will indicate its emergent pole and its latent pole or contrast will be indicated by the point opposite. When the function is an element two points for it may seem supererogatory at first, but it is reasonable to recognize that both can be found and to distinguish between them logically as the pro-element and the anti-element, the former being the one indicated by the coordinates.

Marking both poles of the constructs and one point for each of the elements is the practice that commends itself to common-sense and is certainly the one most frequently followed, but there are also occasions when definite advantages are to be gained from the converse procedure — of mapping both the pro-elements and the anti-elements and only putting in one point for each construct. Suppose, for example, that data, obtained by asking people to say how far they agree or disagree with a series of statements, are recorded as a collective grid, taking the statements as elements and each informant's ratings of them as a construct. Then a point for each construct will represent one person's opinions, and the points for the pro-elements and the anti-elements in its vicinity will show which statements he agrees or disagrees with most.

Globes marked with points for both sets of functions can be used to illustrate the potentialities and also the limitations of rotation. To rotate one manually, simply pick it up and turn it around to any extent in any direction. Notice that the angular distance between all the functions remain constant no matter how it is turned. Evidently any method of estimating the effects of rotation by computer should keep all the angular distance invariant. If it does not apply in exactly the same way to both sets of functions it must produce some misrepresentation of the evidence. If it does, its effect will simply be to bring different aspects of the phenomena into view.

Of course there are other possibilities for using the polar coordinates to map distributions of functions. A perspex globe is particularly useful for displaying which points are opposite or nearly opposite each other diametrically. Another possibility is to use the methods of projection developed by Mercator, Aitoff and others to plot polar coordinates on a flat surface. The relative advantages and disadvantages of different methods of this kind need to be considered when their use is contemplated; none are completely unobjectionable. Yet maps of the earth's surface are so familiar to everyone that analagous ones can be used to explain complicated psychological relationships to people without expert knowledge.

8.8 The rest of the output

An option is available to obtain the tables of residual deviations after the extraction of the components specified in full. That is to say, if g components are being specified the complete series of tables, $D(1)$ to $D(g)$ can be listed. They follow the specifications in sets of three: $D(1)$, $D(2)$ and $D(3)$ after the polar coordinates of the first three components; $D(4)$, $D(5)$ and $D(6)$ after the specifications of the next three, and so on. terminating with $D(k)$ if not before.

Another option can be used to obtain some information about later components, namely their element vectors. The two options may be useful when only a few components are being specified in full. The entries in $D(g)$ will show which points in the grid are overlooked by the imcomplete analysis and the element vectors should show which of the components are most relevant to them. The results might indicate that it would be advisable to repeat the analysis stipulating full specifications for more components.

In practice these options are seldom taken. Most users are satisfied with the customary specifications for the first three components. Those who set other limits are inclined to set them high, preferring to obtain too many results than too few. Moreover, some information about the tables of residuals can be extracted from the specifications given for the components: the values of $V_i(g)$ and $S_j(g)$ give their sums of squares by row and by column. There remain only a few occasions when the output from the options may be of interest.

The output concludes with four tables recording the cosines and angular distances which measure the associations between the constructs and the elements. and between each element and every other one. Their mathematical definitions are the same as those of the correlations and angular distances between the constructs, which are recorded at the beginning of the output (see Section 7.7). The reason why these scales can be used to measure the associations between both kinds of function was explained in Section 8.2. The reasoning may be unfamiliar but the scales are not; presumably there is no need for any explanation of them here.

The user may find it convenient to set out all the correlations and cosines between the functions, or all the angular distances between them, in a single table. Table 9.8 serves as an example.

9

AN EXAMPLE OF THE FINAL PROCESSES IN THE ANALYSIS OF A SINGLE GRID

(with an appendix by *J. B. Watson*)

9.1 The data

The grid which serves as an example of the concluding stages of the analysis was originally reported by J. P. Watson (1970). He obtained it from a girl of seventeen who was admitted to a psychiatric unit as an in-patient when she became anxious and depressed and took to cutting herself on her arms and face. On the basis of her statements and of observations of her behaviour in hospital Watson listed ten situations which he thought might be related to her self-mutilation, and ten possible consequences of being in these situations. Taking each consequence in turn he asked her to rank the situations from the most to the least likely to evoke it. Thus he obtained a grid where the situations are the elements and the consequences the constructs. Table 9.1 lists them and gives the grid.

The main points to consider in attempting to interpret the results from such a grid are the relationships revealed between the functions. Are some particularly closely associated? Or are some strongly contrasted? Are expectations concerning the relationships confirmed? Or have unexpected associations appeared or expected associations failed to appear?

The points to be found in this grid will be examined systematically before any attempt at interpretation is made. The output first shows the relationships between the constructs, expressed as correlations; then those between the elements, expressed as interelement distances; then the relationships between the two sets of functions, defined by reference to the components, and finally in terms of cosines and angular distances.

9.2 Preliminary results

The correlations between the constructs are given in Table 9.2. The first, 'feeling inclined to cut herself', is most clearly associated with thinking people are unfriendly and feeling depressed, and most strongly contrasted with feeling cheerful and likely to be helped in the long run. This is an easily intelligible connection. But there are others which are distinctly perplexing; feeling more grown-up is not contrasted directly with feeling more like a child; it is closely associated with thinking people are unfriendly and feeling depressed. And why is feeling lonely so closely associated with feeling cheerful, while its association with presumably unpleasant feelings — wanting to cut herself, thinking people are unfriendly, feeling depressed, angry and scared — are slightly negative?

Perhaps these problems should be formulated differently. A girl whose prospects

TABLE 9.1   (a) The elements and the constructs concerned

Elements (situations)	Constructs (possible consequences)
A Wanting to talk to someone and being unable to	Likely to:
	1 Make me cut myself
B Having the same thoughts for a long time	2 Make me think people are unfriendly
C Being in a crowd	3 Make me feel depressed
D Seeing G.	4 Make me feel angry
E Being at home	5 Make me feel scared
F Being in hospital	6 Make me feel more grown-up
G Being with my mother	7 Make me feel more like a child
H Being with Dr. W.	8 Make me feel lonely
I Being with Mrs. M.	9 Help me in the long run
J Being with my father	10 Make me feel cheerful

(b) The rankings of the elements in terms of the constructs

Constructs	A	B	C	D	E	F	G	H	I	J
1	2	1	3	6	4	5	7	8	10	9
2	1	3	6	2	4	7	5	8	10	9
3	2	5	3	1	4	6	7	10	9	8
4	1	2	4	3	7	6	10	5	8	9
5	2	3	5	1	9	7	10	4	6	8
6	5	4	7	1	2	6	3	9	10	8
7	2	7	5	9	1	8	3	6	10	4
8	9	8	7	1	5	6	4	2	10	3
9	5	9	10	1	8	2	6	3	4	7
10	9	8	10	1	5	6	3	4	7	2

of growing up are associated with feeling depressed and thinking people are unfriendly and who finds cheerfulness linked with loneliness is presumably in a perplexing predicament herself.

Table 9.3 gives the sums and sums of squares for the elements; the latter are also listed as percentages of the total variation. The relative salience of the elements can be inferred. Evidently the most important ones are being with the boyfriend, G., being with the psychiatric social worker, Mrs. M., and wanting to talk to someone and being unable to. The least salient is being in hospital, which is evidently seen as a rather neutral state.

The elements concerned with home life are found close together in the construct-space (see Table 9.4). Being at home and being with mother are the closest. Being with father is near being with mother but further away from being at home. The elements concerned with psychiatric treatment, being with Dr. W., being with Mrs. M. and being in hospital, are also close together. And hospital situations are associated to some extent with home situations. Being with Dr. W. is close to being with father but being with Mrs. M. is not near being with mother, and neither Dr. W. nor Mrs. M. is near being at home. The first three elements, wanting to talk but being unable to, having the same thoughts for a long time and being in a crowd, form a separate cluster. The association of the third with the first two seems to

TABLE 9.2   Correlations between the constructs

Construct	2	3	4	5	6	7	8	9	10
Likely to:									
1 Make me cut myself	0.770	0.779	0.733	0.370	0.503	0.321	−0.345	−0.442	−0.600
2 Make me think people are unfriendly		0.842	0.673	0.479	0.818	0.333	−0.006	−0.067	−0.127
3 Make me feel depressed			0.636	0.467	0.709	0.188	−0.055	−0.103	−0.212
4 Make me feel angry				0.891	0.224	−0.103	−0.224	0.018	−0.479
5 Make me feel scared					0.067	−0.394	−0.067	0.285	−0.224
6 Make me feel more grown-up						0.297	0.297	−0.018	0.273
7 Make me feel more like a child							0.055	−0.515	−0.067
8 Make me feel lonely								0.345	0.855
9 Help me in the long run									0.479
10 Make me feel cheerful									

TABLE 9.3   Measurements of the elements

Element	Total	Sum of squares	As percentage
A	17.0	114.5	13.9
B	5.0	74.5	9.0
C	−5.0	60.5	7.3
D	29.0	152.5	18.5
E	6.0	60.5	7.3
F	−4.0	24.5	3.0
G	−3.0	66.5	8.1
H	−4.0	68.5	8.3
I	−29.0	124.5	15.1
J	−12.0	78.5	9.5

imply that the girl does not feel absorbed in a crowd, but isolated and alienated from it. Being with her boyfriend is remote from every other situation; it is even almost the most remote from being in hospital, which, lying near the centre-point of the dispersion, is not particularly far from any of the other elements.

The methods used to obtain the results reported in this section are described in Section 7.7 and 7.9.

9.3 Psychological contents of the first three components, considered separately

The first three components have remarkably large latent roots, accounting in all for 87 per cent, of the observed variation (see Table 9.5), while the roots of the remaining components are proportionately small. The loadings of the constructs and the elements, listed in Table 9.6, indicate the psychological contents of the three major components. The first connects the most notable properties of the table of correlations between the constructs and the table of distances between the elements (Table 9.2 and 9.4). Wanting to talk and being unable to and having the same thoughts for a long time (elements A and B) go with wanting to cut herself, thinking people are unfriendly and feeling depressed and angry (constructs 1, 2, 3 and 4). Together they define the positive pole of the component. At the opposite pole, being with mother and father, Mrs. M. or Dr. W. (elements G. H. I and J) are associated with feeling cheerful and being helped in the long run, though also lonely (constructs 8, 9 and 10). The component indicates that the girl feels cheerful in dependent situations, with her parents or the people responsible for her psychiatric treatment; the alternative involves a multitude of afflictions.

The element 'seeing G.' contributes 56 per cent. of the variation along the axis of the second component and its preeminence determines the direction of the axis. The German word *massgebend* is particularly apt for expressing this power found occasionally in a single function to set a scale for measuring all the others. In relation to being with G., the negative values assigned to being in a crowd or with Mrs. M. and the positive values assigned to being alone, feeling more grown-up and more cheerful are not difficult to appreciate.

Since principal components are strictly orthogonal to one another it seems reasonable to suppose that the girl's feelings about her boyfriend do not affect her

TABLE 9.4   Distances between the elements (standardized)

Element	B	C	D	E	F	G	H	I	J	
A Wanting to talk	0.57	0.69	1.12	0.91	0.98	1.23	1.23	1.39	1.38	A
B Having the same thoughts		0.49	1.10	0.84	0.83	1.11	1.10	1.20	1.22	B
C Being in a crowd			1.27	0.76	0.78	1.01	1.03	1.06	1.03	C
D Seeing G.				1.15	0.97	1.18	1.10	1.48	1.29	D
E Being at home					0.82	0.48	1.06	1.27	0.84	E
F Being in hospital						0.73	0.62	0.70	0.75	F
G Being with my mother							0.87	1.08	0.55	G
H Being with Dr. W.								0.78	0.59	H
I Being with Mrs. M.									0.84	I
J Being with my father										

TABLE 9.5 Latent roots of the principal components of the variation recorded in Table 9.1(b)

Component	Root	As percentage
1	351.87	42.65
2	194.85	23.62
3	169.28	20.52
4	45.25	5.48
5	31.31	3.80
6	16.91	2.05
7	12.77	1.55
8	2.12	0.26
9	0.62	0.08

inclination towards self-mutilation very much one way or the other. Construct 1 accounts for less than 2 per cent. of the variation on the axis of the second component, and element 4 less than 5 per cent. of the variation of the first.

The third component reveals a contrast which is similar to the first although it involves a different set of functions. Being at home with mother is associated with feeling more like a child and contrasted with feeling scared and angry; it is also

TABLE 9.6 Loadings of the constructs and elements on the first three components

	Construct	Components		
		1	2	3
1	Wanting to cut myself	8.41	−1.63	−1.51
2	Thinking people unfriendly	8.15	3.23	−1.22
3	Feeling depressed	7.87	2.47	−0.55
4	Angry	7.69	−0.26	4.19
5	Scared	5.38	1.18	6.63
6	More grown-up	5.28	5.69	−3.51
7	More like a child	2.12	−0.32	−7.59
8	Lonely	−2.58	7.59	−1.06
9	Helpful in the long run	−2.34	5.08	5.64
10	Cheerful	−4.49	7.64	−0.92

	Element	Components		
		1	2	3
A	Unable to talk	9.64	−2.05	0.83
B	Preoccupied	7.10	−3.07	0.96
C	In a crowd	4.19	−5.49	−1.01
D	With G.	4.12	10.46	4.73
E	At home	1.95	1.13	−7.37
F	In hospital	−1.94	−0.03	2.37
G	With mother	−3.98	2.69	−6.10
H	With Dr. W.	−5.65	1.25	3.57
I	With Mrs. M.	−7.59	−5.58	4.99
J	With father	−7.82	0.69	−2.86

contrasted with the more external relationships of being with Mrs. M., her boyfriend or Dr. W., and is recognized as not being helpful in the long run.

9.4 Two-dimensional diagrams for showing relationships between elements and constructs

The methods used in the construction of Figure 9.1 were explained in Chapter 8, particularly in Section 8.2. The figure represents the plane of the first two components as a cross-section of the construct-space. The elements are shown as points and the constructs as axes projected onto the surface, their poles being marked around the circumference of a circle drawn with a diameter wide enough to enclose all the elements.

If the opposite poles for construct 1, for instance, were connected by a line and perpendiculars were dropped onto it from the points locating the elements, they would be found to fall in the alphabetical order A to J, which approximates to the order actually assigned to them on construct 1. And if construct 6 were treated similarly, the ordering of the elements would be found to be D, A, B, E, G, C, F, H, J, I, which approximates to the actual order assigned to them on that construct. The figure as a whole gives the closest approximation obtainable with a two-dimensional diagram to the actual orders assigned to all the elements on all the constructs.

The figure draws attention to the isolation as well as the prominance of element D, 'seeing G.', which is the most salient element in the grid. It does not contribute

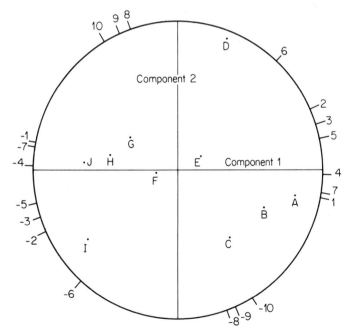

FIGURE 9.1   Composite diagram for components 1 and 2

116

much to the variation along the axis of the first component, which relates to feelings of anxiety, depression and inclination towards self-mutilation, and shows large differences between some elements and others; it lies near the positive pole of the second axis, well away from the rest of the elements. The constructs which distinguish it from them are 6, 8, 9 and 10 — feeling more grown-up, lonely, likely to benefit in the long run and cheerful. There is not much variation between the rest of the elements in this respect.

Most of these points have already been noticed in examining the loadings of the elements and the constructs on the two components separately. The figure does not in fact contribute much to our understanding of the relationships between the two sets of functions.

There is more to be learned from Figure 9.2, which shows the dispersion of the elements in the cross-section of the construct-space defined by the first and third components. It has been constructed like the previous one but is enclosed in a rectangular frame for a change. Construct 8, 'make me feel lonely', has been omitted because it does not contribute much to the variation in this section — over 90 per cent. of its variation is outside (70 per cent. absorbed in component 2).

Component 3 separates situations that diminish the inclination towards self-mutilation, reduce impressions that other people are unfriendly, relieve feelings of depression and promote cheerfulness into two sets: being with mother and father, and being in hospital receiving treatment from Dr. W. and Mrs. M. Both sets have a contrasted element. Being with mother or father is contrasted with seeing G., and Watson's observation (1970) that her parents disapproved of their daughter's association with G. may explain why she feels angry and scared when with him. Being in hospital is contrasted with being at home; and this relationship between the elements explains why feeling more grown-up is correlated positively ($r = 0.3$) instead of being contrasted with feeling more like a child. When she is in hospital, and particularly when she is with Mrs. M., she feels neither grown-up nor like a child; when she is at home she can feel both more childish and more grown-up (see rows 6 and 7 in the original grid, Table 9.1).

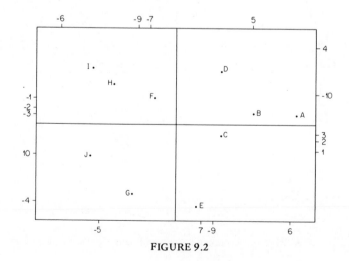

FIGURE 9.2

9.5 Representing relationships between constructs and elements in three dimensions

The dispersion of all the functions of a grid in the space of its three major components can be represented diagramatically using the polar coordinates listed in the output from INGRID 72. Explanations of how these terms are calculated and how they can be used have already been given in Chapter 5, Section 5,11, and Chapter 8, Sections 8.2 and 8.7; a further description of them from the graphic point of view is added here in case it is needed.

As can be seen from Figure 9.1, angular relationships exist between all pairs of functions in a grid. Thus lines can be drawn from the centre of the circle there to any point, be it for an element within the circumference or for a construct on the circumference; and any two of them will form an angle at the point of origin.

In different diagrams the angle may not appear the same. For instance, the angle between constructs 1 and 7 appears much wider in Figure 9.2 than it did in Figure 9.1. There is no reason why it should appear the same in both figures any more than that a house should appear the same from the side as it does from the front. The diagrams represent different two-dimensional aspects of the dispersion of the functions in the k-dimensional component-space.

If the functions are plotted in the space of the first three components a more accurate representation of the angular distances between them can be obtained. Suppose the treatment applied to the constructs in Figure 9.1 is also applied to the elements. The axis for an element is obtained from its loadings and then marked by diametrically opposite points on the circumference of the circle. Element A, for instance, instead of being placed at a point within the circle, is marked on its circumference by poles which would come very close to those of construct 1. Such a pair of poles would indicate the direction in which an element deviates from the origin but not its distance. Their relationship to it might perhaps be conveyed by calling them the pro-element and the anti-element.

After this treatment Figure 9.1 simply consists of the poles of the axes of the constructs and the elements marked round the perimeter of an empty circle. Information about the relative salience of the elements is lost, but the advantage gained is that the method can be extended to combine the results for three components. Then the surface of a sphere takes the place of the circumference of a circle, and the poles of the axes for the elements and the constructs can be mapped on it from their polar coordinates.

A plain globe of the kind supplied for geographers can be used as a base. The poles can be marked with adhesive labels. If both poles of every axis were marked the surface would finally be covered with twice as many labels as the number of constructs and elements together. In Figure 9.3 the pro-elements have been marked but not the anti-elements, in accordance with the suggestion in Section 8.7 that one point is usually enough for the elements if both are marked for the constructs, or vice versa.

A slightly different way of regarding the relationship between the two dispersions is by visualizing the elements as scattered outside the hypersphere where the constructs are located instead of being enclosed within it. The distribution of the constructs can be imagined as rather like that of the population of the earth, congregating more closely in some areas than others, and the distribution of the

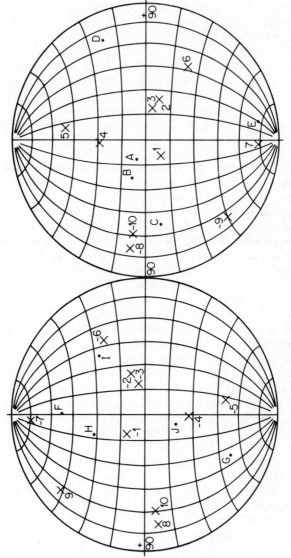

FIGURE 9.3   Composite diagram for components 1, 2 and 3

elements like the scatter of the stars, apparently forming constellations in the heavens above. This model is more appropriate when there are more constructs than elements and they have been normalized, for then the elements are bound to appear on the whole more salient than the constructs while the constructs will all be equidistant from the centre of a hypersphere and therefore scattered on its surface.

The compass bearings of the stars in the sky can be used to map their constellations on a celestial globe. For example, a line extending from the centre of the earth to the north star would intersect the surface of the earth at a point near the north pole, so the position of the north star would be marked there on the celestial globe.

One globe is enough to show which stars are overhead particular points on the earth's surface at a particular time. The constellations can be projected onto the surface of a terrestrial globe, making a celestial and a terrestrial globe rolled into one — provided that the coordinate system used applies to both. Figure 9.3 is constructed in this way. The distributions of the constructs and the elements are mapped on the same surface using the common coordinate system given by the components. The analogy also suggests what reservations are involved when angular distances are abstracted from linear ones.

The configuration of the results from the example is changed in several respects by the inclusion of the third component in Figure 9.3. Constructs 1, 2, 3, 4, 5 and 7, which appeared in Figure 9.1 in the form of a cluster near the positive pole of the first component, are now found dispersed along the axis of the third, with 5 near the positive pole and 7 even nearer the negative. Similarly, construct 9 swings away from 8 and 10 and elements F, H, J and G stretch out.

The negative pole of the third component points to the childish security of being at home with mother, but the situations and feelings that one might expect to find contrasted with it and associated with getting away from home, such as seeing G. and feeling more grown-up, do not appear at the opposite pole. Their absence from this area suggests some disturbance and disorientation of the construct system, especially as what does appear there — feeling scared, being in hospital and likely to help her in the long run — seems to indicate a need for psychiatric help.

Like Figure 9.1 the diagram also illuminates associations between particular elements and constructs, whether they fall in line with any component or not. For instance, the association between J and −4 contrasts being with father to feeling angry; that between I and −6 contrasts being with Mrs. M. to feeling more grown-up; and that of C with −10 contrasts being in a crowd to feeling cheerful.

9.6 A minor dimension of variation

The fact that the three major components account for a very high proportion (86.8 per cent.) of the total variation recorded in the grid does not warrant the conclusion that the rest is without any psychological interest. It is unusual to find an examination of the minor components of a grid totally unrewarding.

Component 6 appears particularly interesting on account of its connection with the presenting symptom in the case, self-mutilation. The construct which refers directly to the symptom is the first. The breakdown of its total variation by component is shown in the third column of Table 9.7. Although its loading on the sixth component is a good deal smaller than on the first it contributes a

TABLE 9.7

Component	(2) Latent root	(3) Variation contributed by construct 1	(3) as percentage of (2)
1	357.87	70.65	20.1
2	194.85	2.65	1.4
3	169.28	2.28	1.3
4	45.25	0.03	0.1
5	31.31	0.43	1.4
6	16.91	4.07	24.0
7	12.77	2.01	15.7
8	2.12	0.34	16.0
9	0.62	0.06	9.0

proportionally larger amount to the variation along the axis of the sixth, where the total is smaller still.

The axis is shown as vertical line in Figure 9.4, with the constructs distributed along it on the right and the elements on the left. In terms of the constructs it indicates a bipolar contrast between self-mutilation and depression. This suggests

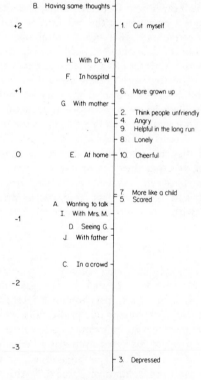

FIGURE 9.4

that self-mutilation offers a temporary relief from depression. Treatment is placed in delemma: if it is directed towards relieving depression it may be liable to produce a swing towards self-mutilation, but if it is intended to discourage self-mutilation it may be liable to induce depression. The locations of the elements on the axis suggest that Dr. W.'s treatment tends to have the former effects and Mrs. M.'s the latter. The locations of 'being at home' and 'feeling cheerful' at the centre-point of the axis only indicate that these functions have no influence on the variation along it.

As there is so much of interest in the variation along this minor axis it may be advisable to recall that the first component, which of course is strictly orthogonal and accounts for 43 per cent. of the observed variation, combines both feelings and shows that both treatments tend to alleviate them; and to note that the sixth, where these distinctions between them become apparent, only accounts for 4 per cent.

9.7 The complete set of angular relationships between the functions

The angular relationships between the functions are only shown approximately in Figure 9.3 because their dispersion in the component-space is only approximately three-dimensional. To calculate their relationships exactly the method given in Section 7.2 needs to be used. INGRID 72 concludes by giving tables of the relationships between the constructs and the elements and of the elements with one another first in terms of cosines and then of degrees.

As the cosines are mathematically equivalent to correlations the output from this part of the program can be combined with the previous listing of the correlations between the constructs (see Table 9.2) to make a table where all the relationships between pairs of functions are expressed in comparable terms. Table 9.8 gives the complete array. The relationships could be tabulated similarly in degrees if desired. (The alternatives are discussed in Section 8.2.)

Such a table may be examined to check impressions derived from other aspects of the results and to see which functions are most closely associated and which are most completely contrasted, taking all the functions of both kinds into consideration.

For example, the associations between particular elements and constructs noticed in Figure 9.3 are measured in Table 9.8 as

Construct with element		Cosine
4	J	−0.84
6	I	−0.81
10	C	−0.85

Again, the entries in the top row of the table show that construct 1, 'likely to make me want to cut myself', is most closely associated with elements A and B, 'wanting to talk to someone and being unable to', and 'having the same thoughts for a long time'. Construct 6, 'likely to make me feel more grown-up', is most closely associated with 2, 'likely to make me think people are unfriendly', while this in turn is most closely associated with 3, 'likely to make me feel depressed'.

TABLE 9.8 The direction cosines or correlations between the elements and the constructs in the example

	Constructs									Elements									
	2	3	4	5	6	7	8	9	10	A	B	C	D	E	F	G	H	I	J
1	0.77	0.71	0.73	0.37	0.50	0.32	-0.34	-0.44	-0.60	0.81	0.87	0.65	0.08	0.36	-0.32	-0.39	-0.69	-0.66	-0.80
2		0.84	0.67	0.48	0.82	0.33	-0.01	-0.07	-0.13	0.77	0.60	0.17	0.54	0.41	-0.42	-0.17	-0.62	-0.82	-0.75
3			0.64	0.47	0.71	0.19	-0.06	-0.10	-0.21	0.71	0.52	0.37	0.54	0.33	-0.32	-0.30	-0.73	-0.70	-0.72
4				0.89	0.22	-0.10	-0.22	0.02	-0.48	0.82	0.77	0.44	0.41	-0.25	-0.21	-0.82	-0.24	-0.46	-0.84
5					0.06	-0.40	-0.07	0.28	-0.22	0.59	0.54	0.17	0.58	-0.56	-0.12	-0.85	0.05	-0.19	-0.66
6						0.30	0.30	-0.02	0.27	0.31	0.27	-0.13	0.60	0.63	-0.28	0.29	-0.62	-0.81	-0.43
7							0.06	-0.52	-0.07	0.32	-0.01	0.21	-0.33	0.12	-0.62	0.46	-0.33	-0.58	0.12
8								0.34	0.86	-0.47	-0.51	-0.59	0.55	0.83	-0.09	0.40	0.46	-0.40	0.45
9									0.48	-0.17	-0.47	-0.75	0.59	-0.53	0.65	-0.10	0.49	0.22	-0.04
10										-0.62	-0.67	-0.85	0.51	0.08	0.05	0.58	0.43	-0.13	0.57
A											0.70	0.52	0.13	0.13	-0.34	-0.56	-0.54	-0.48	-0.83
B												0.68	0.02	0.03	-0.33	-0.59	-0.56	-0.35	-0.78
C													-0.43	0.13	-0.36	-0.48	-0.51	-0.12	-0.41
D														-0.16	0.03	-0.17	0.00	-0.45	-0.35
E															-0.16	0.66	-0.60	-0.64	0.06
F																-0.07	0.28	0.54	-0.01
G																	-0.03	-0.12	0.62
H																		0.43	0.57
I																			0.36

The alternatives the girl sees open to her are revealed by contrasts between some functions and others. The most direct alternatives to J, being with father, are 4, feeling angry, A, being unable to talk to anyone, and I, feeling inclined to cut herself. The alternatives to G, being with mother, are 4, feeling angry, and 5, feeling scared. The alternatives to 7, feeling more like a child, are F, being in hospital, and I, being with Mrs. M.; but being with Mrs. M. also presents the most distinct contrast to 6, feeling more grown-up. The alternatives to E, being at home, are H and I, being with Dr. W. or Mrs. M.

9.8 Conclusions

The girl appears from her grid to be caught in a trap. It shows no way for her to achieve normal maturation into adulthood. She may turn towards her home and her parents, or towards the hospital and her therapists; otherwise she is faced with anger, fear, depression and self-mutilation. Being with her boyfriend does not offer her an escape.

9.9 Apendix: psychological aspects (contributed by J. P. WATSON)

The patient who completed the grid discussed by Slater was treated by me in hospital between May 1968 and May 1969. She was a petite, quiet, not unattractive girl with parents who were suspicious of doctors and hospitals and quite unpredictable towards her. There was a good deal of maladjustment in the family, for two sisters and the mother had received psychiatric treatment in the past and a brother had had criminal connections. The one family member of whom the patient spoke warmly was a fourth sibling, a married sister three years her senior.

The patient had attended school until aged fifteen, but had been a barely average scholar and lonely and unhappy at school, feeling herself cut off from her fellows. She had shown some promise as an apprentice hairdresser after leaving school but stopped work after a year when emotional turmoil and self-mutilation mounted as she became progressively more involved with a twenty-one year-old man. She was admitted to another hospital for a short period, and the self-mutilation was both seen as the major problem and ascribed to difficulties in the relationship with the boyfriend, of whom her parents disapproved.

Repeated clinical observation in our unit suggested, however, that family and other interpersonal problems were in their way as serious as the self-mutilation, and that a doctor who was temporarily able to put the behaviour disorder aside could elicit symptoms of depression, free-floating and phobic anxiety, depersonalization, and symptoms with obsessional and hypochondriacal features. There were multiple neurotic symptoms as well as behavioural characteristics typical of patients often diagnosed as hysterical personality disorder or psychopathy. As is often the case with such patients, the behaviour was usually of more concern to unit staff than the distress. It is understandable that a nurse should act when a patient says that they have hidden a razor-blade and might cut themselves, or breaks a window and runs off with a piece of broken glass, or presents herself albeit apologetically with a bleeding arm requiring ten to twenty sutures, and that the nurse should categorize the patient as attention-seeking, manipulative and psychopathic when such things occur repeatedly despite strenuous efforts by all staff.

Self-mutilation occurred at roughly fortnightly intervals throughout the period in hospital, and neither admission alone, nor a variety of drugs, nor once-weekly psychotherapy from me, prevented her from injuring herself. A repertory grid investigation was carried out over the latter part of the hospitalization to try and classify the difficult clinical problem. My subjective impression at the time was that cutting behaviour had become related to my psychotherapeutic activity, though occurring on average as frequently as before. Some support for this notion may perhaps be derived from her career after I left the unit at the end of May 1969, when she was still an in-patient. During June 1969 the patient broke unnumerable windows, slashed herself repeatedly, and attacked her legs and abdomen as well as her arms. She was transferred to a mental hospital at the end of June and the self-mutilation abruptly ceased.

Self-mutilant behaviour is often associated with very difficult clinical problems and this patient posed many of them — it is a problem for the psychiatrist if he is unable to stop maladaptive behaviour while feeling that this should be his primary aim. The psychiatrist may also have to deal with the strong feelings among both staff and other patients which recurrent self-mutilation usually evokes. Further, it is not always clear why treatment is unsuccesful, which is frequently the case, or why the behaviour occurs. The tendency, already alluded to, for staff dealing with self-mutilant patients to attend to their deviant behaviour rather than to their feelings or subjective experiences may be related to a tendency to regard self-mutilant acts as responses to environmental events. A doctor is particularly likely to misunderstand his patients if problems such as these arise, for they tend to close his mind to important aspects of his patient's difficulties. In any case, the general nature of the doctor—patient relationship and of the relationships between a psychiatrist and his colleagues and patients are so complex that no one person can hope to comprehend all their facets.

Repertory grid method may help to clarify clinical problems such as these, for they may be used to check hypotheses made about individual patients, to bring neglected aspects of problems into perspective, to obtain information often not easily obtained in other ways and to gain unsuspected but useful information. The results of a repertory grid investigation may make some clinical hypotheses seem less likely and others more probable, according to the confidence with which psychological inferences can be made from the data. I think that grids can be valid and reliable in this general sense, but not if the terms 'validity' and 'reliability' have their traditional meanings. As Slater has pointed out, the postulates of probability theory cannot be readily adapted to grids.

Several items of clinically useful information were obtained from the grid investigation discussed in Slater's paper (Watson, 1970).

(a) The patient percieved certain subjective states ('having the same thoughts in my head for a long time' and 'wanting to talk to someone and being unable to') as more likely antecedents of her self-mutilation than the other 'elements' she was asked to judge. This view of her behaviour differed from mine, and implies either that I had ignored an important aspect of her problem, or that she was right and I was wrong, or that I was right and she had given misleading test responses. The latter two possibilities are sterile.

(b) The test results implied that the patient's previous boyfriend, G., was of much greater current importance in her daily thinking and experience than I had supposed. The relationship had been discussed at length in the early weeks of psychotherapy and I had clearly deceived myself about the effectiveness of these discussions.

(c) The patient's relationships with her parents and with me were, according to the grids, not related to the self-mutilation as I had previously thought. Our initial formulation had been that certain environmental events, such as interactions with her mother having certain characteristics, provoked self-mutilation. As stated, this is clearly too simple to be likely to be adequate, but it is of the same type as many formulations of behaviour made in ordinary life and by psychiatrists, such as 'I got drunk because I failed my exam' or 'she is depressed because her father has died'. In the example, the grids suggested that certain unpleasant experiential states were more probably the immediate antecedents of self-mutilation than environmental events. These might, of course, have evoked the relative subjective states. The clinical result was to suggest the value of encouraging the patient to learn to tolerate unpleasant experiences better, so that she might feel less drive to end them by 'acting-out'.

The clinical approach involved here is conveniently described in personal construct terms. Psychiatric treatment may imply that its recipient needs to change his construing system; this is likely to be hindered if the psychiatrist's construing system does not comprehend the most important aspects of his patient's system. In other words, a psychiatrist requires personal constructs whose ranges of convenience can comprehend their patient's problems. The same principle doubtless applies to satisfactory personal relationships in everyday life.

10

THE RELIABILITY AND SIGNIFICANCE OF
A GRID

10.1 Statistical and logical theories of probability pertinent to grids

It may seem perfectly reasonable to stipulate that a grid should be reliable and significant if important decisions depend on the results from it — for instance, if the diagnosis or treatment of a patient is concerned. Moreover, the array of data in a grid, giving an informant's evaluations of a set of elements in terms of a set of constructs, is the same in its general form and in most of its mathematical properties as a two-way table giving the scores of a group of subjects on a battery of tests — the subjects correspond to the elements and the tests to the constructs. So why should not the reliability and significance of both kinds of array be investigated by the same methods?

They cannot. The reason is that the theory from which psychometric methods for measuring reliability and significance are derived assumes that samples can be drawn at random from an objectively defined population. The assumption can be satisfied by the nomothetic data in table of test scores, but not by the idiographic data in a grid.

A typical occasion for giving a battery of tests to a group of subjects occurs when their scores are needed for allocating them to different educational courses or different duties in a large organization. Before the tests can be put to such use their reliability and significance should be investigated thoroughly. An applicant's scores should characterize him as a person. not just record a temporary state of mind. The norms for the courses or the duties should be known, and the differences between them shown to be highly significant. Then it would be possible to estimate from the candidate's scores what is his probability of adapting successfully to each of the alternative situations and hence to decide which is the best one for him (or whether he is unfitted for any).

To find out whether a test in the battery is reliable it should be given twice, with a reasonable time in between, to a random sample of the population from which the applicants come. The correlation between the scores on the two occasions, which is the measure of the test's reliability, will be an unbiased estimate of the correlation in the population, provided the sample is a random one. For the same reason the norms for the courses or duties should be obtained from random samples of the people who have adapted successfully to them, if not from the records of everyone who has done so; the proof that the norms vary significantly would be invalid if they were obtained from unrepresentative samples. All these investigations need to be completed before the battery of tests is ready for use in the selection procedure.

No such preliminary investigations can be carried out when a grid is constructed specially for one informant on one occasion. The setting is likely to be a clinic, the interviewer a consultant and the informant a patient who presents some psychological problem. The clinician and the patient will collaborate in constructing the grid, with the problem as its focus of interest. Eliciting the elements and the constructs and filling in the grid will be one continuous process. The results will be wanted as soon as possible. The clinician may discuss them with the patient and agree on their interpretation.

Since an idiographic grid refers to a population of elements which cannot be defined objectively or sampled at random the data for assessing its reliability and significance could not be obtained even if there was time to spare. But besides, there is not the same need for an assessment. The grid does not serve the same purposes as a battery of tests. Its primary interest is in what it shows directly — the informant's state of mind at the time of interview. Its predictive value for estimating what is to be expected in another case or on another occasion may not need to be considered, and constructs which register changes in mental states may be more suitable for inclusion in it than ones that record stable personality traits. Thus the criteria of significance and reliability proposed by statistical theory are inappropriate for it as well as inapplicable.

Must the evidence be discarded because it does not conform to statistical canons? We seem to be faced by the dilemma: if the orthodox statistical theory of probability is sound the scientific status of grid technique is questionable, while if the technique is acceptable the theory is called into question. In that case, however, there is no need to question the theory as a whole, but only its application to idiographic grids.

It has long been recognized that statistical methods are limited in their application and that probabilities have often to be estimated on the basis of evidence of other kinds. Suppose, for instance, that someone is charged in court with committing a crime. Statistical evidence that people like him commit similar crimes in similar circumstances will not be admitted in evidence, no matter how high the probability they establish. Only evidence unambiguously proving or disproving the defendent's personal connection with the crime will satisfy the court. While the case is in progress and the evidence is being presented the probability that he will prove innocent fluctuates perceptibly.

Keynes (1921) achieved remarkable success in his original attempt to develop a logical theory of probability which could be extended to such contingencies. But the problem of defining the most reasonable estimate is formidable when the evidence cannot be evaluated by reckoning up instances. Bayesian statistical methods have no advantage over traditional ones in this respect. Both derive estimates of probability from the simple enumeration of instances and are equally incapable of handling the jig-saw of miscellaneous pieces of evidence. Though protracted studies of the subject have been made since (notably by Carnap and others, 1971), no methods have yet been found for quantifying logical probability with the neat mathematical expressions, $p < 0.05$, etc., which have become the hallmarks of acceptable psychological papers.

Even supposing that arguments of a statistical kind can be adduced for expecting that grids are generally reliable provided they satisfy certain conditions — whether mathematical or psychological — the proposition that a particular idiographic grid is reliable in a particular case must remain open to doubt. And it cannot be decided by statistical evidence; what is needed is evidence that is logically relevant. It must establish definite connections between the contents of the grid and what is known about the informant from other sources, or can be verified by further investigation. We should recognize that if such evidence is logically valid it pre-empts the call for statistical tests.

10.2 Properties of arrays of random numbers treated as grids

The null hypothesis that a particular grid is indistinguishable from an array of random numbers may need to be considered in some circumstances. So a program has been developed to study the properties of such arrays. It generates n by m arrays of random numbers and treats them as if they were grids with n constructs and m elements. It applies an abbreviated form of the principal component analysis used by INGRID for experimental grids, omitting the calculation of eigenvectors the functions derived from them. As it generates and analyses grids of random numbers it has been called GRANNY for short. The grids it generates are called 'quasis' here to constrast them with ones obtained experimentally.

The philosophical problem of whether any series of numbers generated mathematically can be truly random has been evaded by delegating the responsibility to the Fortran Function RANF(X). It continues generating a series of numbers as long as required when starting from a seed, which may be any five-digit number. Though the series is predetermined by the seed the user cannot anticipate what it will be before starting it nor tell from its output up to any point how it will continue thereafter. His only advance information is that all real numbers in the range 0 to 1 are equally likely to occur in it in any order. In calling it, he acts more like a gambler taking a chance than a statistician does when he consults a table of random numbers. He can be sure that the processes involved in forming quasis are independent of whatever psychological processes generate the sets of numbers in experimental grids.

At present two versions of GRANNY are available, one for grids with more constructs than elements, with its upper limits set at $n = 50$ and $m = 25$, and a converse one with its upper limits at $n = 25$ and $m = 50$. The real numbers obtained from RANF(X) may be entered as they come or they can be converted into rankings or gradings on any scale, two-point or over, according to choice. The quasis so constructed can be listed if desired. If they are graded they may be normalized or not, as preferred. Up to one hundred quasis can be generated at a time.

The variation within these limits is so wide that publication of a complete set of tables does not appear to be a practical proposition yet. It would also be unnecessarily extravagant to run the program repeatedly to obtain extra sets of quasis with the same specifications. Summaries of the output from sets with given specifications will accordingly be duplicated and made available on personal request. Requests addressed to the author should give the dimensions of the experimental grid to be matched, and may specify the method of sorting to be used. If they ask for a grading they should also state whether normalization is to be applied or not.

Eventually it may prove possible to produce a compendious set of tables which will satisfy most requirements, either by accumulating results or by mathematical generalization. The exact solution of a mathematical problem is often easier to find when a practical method for approximating to it is already available. Moreover, the definition of the problem may be simplified if accumulating results show that some approximations are quite adequate for practical purposes — as following results tend to suggest.

Tables 10.1 and 10.2 illustrate the output from the program. They are derived

TABLE 10.1   Summary of properties of one hundred random 10 by 10 arrays treated as grids graded on a two-point scale, not normalized

| | Range | | Mean | |
	Upper limit	Lower limit		Standard deviation
Bias	0.4561	0.1673	0.3030	0.0635
Variability	1.0371	0.9428	1.0021	0.0225

Size of the latent roots as percentage of:

(a) The total variaton about the construct means

Component 1	41.23	24.22	30.38	3.53
2	28.39	16.88	22.27	2.39
3	21.91	11.23	16.69	2.11
4	16.04	7.72	12.17	1.79
5	12.84	4.74	8.63	1.49
6	7.97	2.42	5.27	1.36
7	5.83	0.60	3.05	1.00
8	3.70	0.27	1.26	0.66
9	0.96	0.00	0.27	0.22

(b) The residual variation after the extraction of the previous roots

2	45.05	24.16	32.09	4.05
3	46.34	23.74	35.37	4.22
4	52.32	27.70	39.87	4.75
5	66.98	33.75	47.13	6.88
6	77.03	38.56	53.89	8.06
7	87.10	48.43	67.17	9.28
8	100.00	51.10	81.73	12.47

(c) Cumulative percentage extracted by increasing numbers of components

Up to 2	67.54	45.39	52.65	4.41
3	78.77	60.39	69.34	3.98
4	89.20	72.50	81.51	3.15
5	95.01	83.16	90.15	2.45
6	98.83	91.50	95.42	1.51
7	99.64	95.88	98.47	0.73
8	100.00	99.04	99.73	0.22

TABLE 2 Frequency distributions

Construct correlations			Angular distance between constructs			Element distances		
Below	Frequency	Proportion	Below	Frequency	Proportion	Below	Frequency	Proportion
-0.85	0	0.0000	5°	4	0.0009	0.1	4	0.0009
-0.75	27	0.0060	15	0	0.0009	0.2	0	0.0009
-0.65	59	0.0191	25	0	0.0009	0.3	0	0.0009
-0.55	93	0.0398	35	0	0.0009	0.4	0	0.0009
-0.45	133	0.0693	45	51	0.0122	0.5	34	0.0084
-0.35	457	0.1709	55	167	0.0493	0.6	0	0.0084
-0.25	81	0.1889	65	188	0.0911	0.7	173	0.0469
-0.15	709	0.3464	75	529	0.2087	0.8	444	0.1456
-0.05	318	0.4171	85	1015	0.4342	0.9	712	0.3038
0.05	669	0.5658	95	669	0.5829	1.0	1043	0.5356
0.15	287	0.6296	105	1027	0.8111	1.1	1040	0.7667
0.25	728	0.7913	115	500	0.9222	1.2	702	0.9227
0.35	85	0.8102	125	171	0.9602	1.3	293	0.9878
0.45	486	0.9182	135	152	0.9940	1.4	53	0.9996
0.55	146	0.9507	145	27	1.0000	1.5	2	1.0000
0.65	88	0.9702						
0.75	79	0.9878	Mean 89.2063			Mean 0.9868		
0.85	51	0.9991	s.d. 20.0444			s.d. 0.1621		
0.95	0	0.9991						
1.05	4	1.0000						

Mean 0.0122
0.3293

note. The frequencies lie within implied limits. Thus 27 constructs' correlations were below −0.75 but not below −0.85. The proportions are cumulative.

from a set of one hundred 10 by 10 quasis graded on a two-point scale without normalization.

The formulae for the measures of bias and variability calculated by INGRID for graded grids can be applied to quasis, and GRANNY lists the values so obtained. The values for the ones examined here are summarized at the beginning of Table 10.1 by giving their means and standard deviations, and the upper and lower limits of their ranges.

GRANNY stores the latent roots of quasis from largest to smallest and lists them in three different ways: firstly, each as a percentage of the total variation about the row means; secondly, each as a percentage of the residual variation after the previous roots have been subtracted; and thirdly, the cumulative percentage of the total variation attributable to increasing numbers of components is listed from the first two up to the last but one. Each of these ways of examining the relative sizes of the latent roots of quasis may conceivably be found suitable for comparison with experimental grids. The results from the present set are summarized in Table 10.1 by giving their ranges, means and standard deviations.

In Table 10.2 the distribution of the correlations between the quasi-constructs is listed just as it appears in the output from GRANNY, and so, too, are the distributions of their angular distances and the standardized linear distances between the quasi-elements.

Table 10.3 provides briefer summaries of the results from using other sorting methods to form one hundred 10 by 10 quasis from the same series of random numbers. Part 1 gives results for two-point, five-point and seven-point scales, normalized; Part 2, results for the same scales, not normalized; and Part 3, results for real numbers, both normalized and not normalized, and for ranking. The real numbers are accepted straight from RANF(X); the results from them are specially appropriate for comparison with the grids of differential change and consensus grids obtained from DELTA and SERIES (see Chapter 11). Ranking and grading on two-point, five-point and seven-point scales are the sorting procedures most frequently used (Chetwynd, 1974).

All the distributions show a well-marked tendency to approximate to the normal form. They are continuous and unimodal, tapering finely towards both extremes. Some are practically symmetrical. The most skewed are naturally those of the minor eigenvalues. In the distribution of the standardized distances between the elements, where the range might have been expected to be greater above its mode than below, the skewness turns out to lie in the opposite direction. The distribution of the correlations between the constructs appears to be distinctly platykurtic, but not so that of the angular distances between them. The distributions for other sorting methods are smoother than for two-point gradings, where some oddities are to be found.

10.3 Differences in rating methods

As already explained, the same series of numbers is obtained from RANF(X) every time the same seed is used. To make the results from different rating methods as closely comparable as possible, the same seed has been used to generate the quasis for each method reported in Table 10.3. Differences between the methods provide the only possible explanation for the differences in the results. Nevertheless they

TABLE 10.3

Part 1   Means and s.d.s of measurements of quasis using different scales

Measurement	Two-point scale normalized		Five-point scale normalized		Seven-point scale normalized	
	Mean	S.d.	Mean	S.d.	Mean	S.d.
Bias	0.3030	0.0635	0.2191	0.0435	0.2077	0.0421
Variability	1.0021	0.0225	0.7051	0.0307	0.6679	0.0298

Relative sizes of successive roots, as proportions:
(a) Of the total variation

Root	1	30.13	3.48	30.14	3.41	30.17	3.59
	2	22.25	2.53	22.53	2.34	22.19	2.34
	3	16.72	2.11	16.82	1.83	16.81	1.77
	4	12.15	1.72	12.20	1.90	12.14	1.84
	5	8.69	1.43	8.30	1.49	8.46	1.43
	6	5.37	1.36	5.44	1.46	5.58	1.53
	7	3.12	0.97	2.86	1.10	3.01	1.17
	8	1.29	0.69	1.36	0.72	1.32	0.69
	9	0.28	0.22	0.34	0.31	0.33	0.30

(b) Of the residual variation

Root	2	31.93	4.04	32.32	3.63	31.83	3.55
	3	35.23	4.22	35.72	4.47	35.47	4.28
	4	39.49	4.51	40.33	6.34	39.72	6.07
	5	46.80	6.63	45.93	6.29	45.93	6.59
	6	53.75	8.04	55.01	8.04	55.22	7.80
	7	67.25	9.31	63.36	10.58	65.12	10.52
	8	81.67	12.56	80.31	12.06	80.76	11.70

Proportional sizes of successive roots cumulatively

Root	2	52.39	4.26	52.67	4.08	52.35	4.06
	3	69.11	3.87	69.49	4.15	69.16	4.09
	4	81.26	3.08	81.69	3.73	81.31	3.72
	5	89.95	2.41	89.99	2.77	89.77	2.86
	6	95.32	1.51	95.43	1.72	95.35	1.72
	7	98.43	0.75	98.30	0.88	98.35	0.86
	8	99.72	0.22	99.66	0.31	99.67	0.30

Construct correlations

	0.0122	0.3293	0.0140	0.3308	0.0147	0.3291

Angular distances between constructs

	89.21	20.04	89.15	20.03	89.10	19.93

Linear distances between elements

	0.9857	0.1686	0.9845	0.1754	0.9846	0.1750

TABLE 10.3   (cont)

Measurement	Two-point scale not normalized		Five-point scale not normalized		Seven-point scale not normalized	
	Mean	S.d.	Mean	S.d.	Mean	S.d.
Bias	0.3030	0.0635	0.2191	0.0453	0.2077	0.0421
Variability	1.0021	0.0225	0.7051	0.0307	0.6679	0.0298

Relative sizes of successive roots, as proportions:
(a) Of the total variation

Root 1	30.38	3.53	31.58	4.00	31.63	3.87
2	22.27	2.39	22.78	2.66	22.62	2.55
3	16.69	2.11	16.67	1.96	16.58	1.89
4	12.17	1.79	11.84	1.97	11.74	1.87
5	8.63	1.49	7.88	1.66	7.94	1.57
6	5.27	1.36	5.03	1.31	5.26	1.42
7	3.05	1.00	2.65	0.99	2.73	1.04
8	1.26	0.66	1.26	0.65	1.21	0.64
9	0.27	0.22	0.30	0.28	0.30	0.28

(b) Of the residual variation

Root 2	32.09	4.05	33.40	4.28	33.18	4.15
3	35.37	4.22	36.76	4.60	36.47	4.49
4	39.87	4.75	41.22	6.23	40.59	5.89
5	47.13	6.88	46.38	5.91	46.06	6.35
6	53.89	8.06	55.07	8.12	56.06	7.79
7	67.17	9.28	63.55	10.47	65.00	10.66
8	81.73	12.47	81.06	11.94	81.00	11.84

Proportional sizes of successive roots cumulatively

Root 2	52.65	4.41	54.36	4.75	54.25	4.60
3	69.34	3.98	71.04	4.43	70.83	4.32
4	81.57	3.15	82.87	3.67	82.57	3.64
5	90.15	2.45	90.75	2.48	90.57	2.61
6	95.42	1.51	95.79	1.54	95.77	1.55
7	98.47	0.73	98.44	0.80	98.49	0.79
8	99.73	0.22	99.70	0.28	99.70	0.28

Construct correlations

	0.0122	0.3293	0.0140	0.3308	0.0147	0.3291

Angular distances between constructs

	89.21	20.04	89.15	20.03	89.10	19.93

Linear distances between elements

	0.9568	0.1621	0.9838	0.1793	0.9838	0.1792

TABLE 10.3 cont. (Part 3)

Measurement	Real numbers normalized		Real numbers not normalized		Ranking	
	Mean	S.d.	Mean	S.d.	Mean	S.d.
Bias	0.1779	0.0363	0.1779	0.0363		
Variability	0.5772	0.0278	0.5772	0.0278		

Relative sizes of successive roots, as proportions:
(a) Of the total variation

		Mean	S.d.	Mean	S.d.	Mean	S.d.
Root	1	30.21	3.47	31.75	3.89	30.46	3.67
	2	22.27	2.38	22.79	2.41	22.33	2.47
	3	16.89	1.79	16.49	2.05	16.91	1.81
	4	12.07	1.87	11.72	1.99	12.07	1.78
	5	8.39	1.51	7.83	1.56	8.20	1.64
	6	5.49	1.43	5.21	1.36	5.47	1.50
	7	2.98	1.13	2.68	1.02	2.99	1.20
	8	1.36	0.69	1.26	0.62	1.28	0.69
	9	0.33	0.30	0.29	0.27	0.30	0.28

(b) Of the residual variation

		Mean	S.d.	Mean	S.d.	Mean	S.d.
Root	2	31.97	3.59	33.49	4.00	32.20	3.96
	3	35.73	4.35	36.46	4.65	36.08	4.79
	4	39.75	6.22	40.73	6.03	40.30	6.07
	5	45.76	6.26	45.76	6.15	45.56	6.42
	6	54.69	8.00	55.81	7.96	55.51	9.02
	7	64.36	10.27	63.97	10.56	66.06	10.01
	8	81.25	11.65	81.98	11.32	81.68	11.65

Proportional sizes of successive roots cumulatively

		Mean	S.d.	Mean	S.d.	Mean	S.d.
Root	2	52.48	4.00	54.53	4.59	52.79	4.45
	3	69.37	4.07	71.02	4.26	69.70	4.55
	4	81.44	3.68	82.74	3.52	81.77	3.97
	5	89.83	2.74	90.57	2.50	89.97	2.85
	6	95.32	1.70	95.77	1.53	95.44	1.81
	7	98.30	0.85	98.45	0.78	98.43	0.85
	8	99.67	0.30	99.71	0.27	99.70	0.28

Construct correlations

Mean	S.d.	Mean	S.d.	Mean	S.d.
0.0143	0.3299	0.0143	0.3299	0.0149	0.3328

Angular distances between constructs

Mean	S.d.	Mean	S.d.	Mean	S.d.
89.13	19.96	89.13	19.96	89.09	20.18

Linear distances between elements

Mean	S.d.	Mean	S.d.	Mean	S.d.
0.9844	0.1759	0.9835	0.1807	0.9853	0.1706

are dependent on the choice of seed and may be peculiar to it; they certainly would not be replicated exactly with another one.

Most of the observed differences are slight and irregular and appear to be negligible for practical purposes. This does not imply that all methods of rating are equally suitable for obtaining any experimental grid — certain methods are well known to be particularly suitable in certain cases — but only that one method may sometimes be as good as another for obtaining a set of results from GRANNY to compare with an experimental grid.

Bias and variability (b and y, see Section 7.3) are the measurements most affected by differences in rating. The results in Table 10.3 seem to indicate that values obtained from grids using different rating methods may not be strictly comparable. In any investigation of individual differences it would therefore seem advisable to get all the informants to use the same rating method, and probably also to match their grids in some other respects, e.g. to use a common set of elements.

Though y is dependent on b when two-point grading is used it is not necessarily a single-valued function of b. The general tendency, which is well marked in the results reported here, is not surprisingly for it to diminish as b increases. When scales with more points are used the dependence of y on b diminishes. The results show that it is practically negligible for the five-point and seven-point scales. The correlations between b and y here are -0.1667 and -0.1107 respectively.

Another interesting difference between the rating methods not revealed in Table 10.3 is to be found in the frequency of occurrence of zero eigenvalues; ten of the two-point quasis have their last latent root identically equal to zero; none of the others do, though there are instances when it is less than 0.005 and is consequently listed as zero correct to two decimal places. There are many ways by which variation may come to be excluded from some dimension of the component-space of an experimental grid; for instance, a pair of elements or a pair of constructs may be located at the same point in it if the informant fails to make any distinction between them. He is most likely to do so if the only distinctions he is free to make are dichotomous — conversely if he ranks the elements he is constrained to distinguish between all of them. Still he may identify two constructs by ranking the elements in the same order on both of them. There are other more complicated and unusual ways by which variation may be eliminated from one dimension, and any of them may be simulated accidentally by GRANNY. Table 10.2 shows that zero distances betweens elements and perfect correlations between constructs have occurred in the two-point quasis, but not enough to account for all ten zero eigenvalues.

10.4 Some notes on differences between experimental grids and quasis

Experimental grids hardly ever resemble arrays of random numbers, even remotely. Undoubtedly the commonest and most conspicuous difference is the relatively large amount of variation associated with the major component. Human construct systems evidently tend to be self-reinforcing; later constructs confirm ones elicited earlier in the experimental situation; and the major axis of the dispersion of the elements in the construct-space is almost always very prominent.

If the first component accounts for an unexpectedly large amount of the variation in his grid, an experimenter may go on to see whether the second accounts for more than expected of the remainder, or alternatively whether the first two together account for an unexpectedly large amount of the total. Either possibility can be considered by referring to the output from GRANNY (see Table 10.3), and so can similar possibilities extending to further dimensions.

The second component often has a well-defined axis, and sometimes so does the third, but sooner or later there is usually a marked drop in the size of the latent roots, followed by several that only differ by proportionately small amounts. This indicates a region in the component-space where the dispersion of the elements is approximately spherical and the orientation of the eigenvectors is less distinct. It looks the obvious stopping point for an incomplete interpretation of the data. The previous components exhibit the most conspicuous features of the grid and there are no special reasons for paying attention to any particular one among those that follow.

But if some parts of the variation prove too large to be consistent with the null hypothesis mathematical arguments cannot reinstate it later or prove that the residual variation beyond a certain point must be dismissed as devoid of psychological interest. If the variation is unexpectedly large in some respects there must, conversely, be other respects in which it is unexpectedly small. The experimenter is confronted with the problem of why variation has been considerably restricted or even entirely suppressed in a certain dimension. He may choose to ignore such problems but should not assume that they do not arise. Their explanation may sometimes be simple, even trivial, but seldom if ever psychologically meaningless.

Statistical arguments might perhaps be applied more effectively if some way could be found for obtaining an unbiased estimate of error variance from an experimental grid, but no method tried so far has proved adequate. One proposed for monitoring changes in a patient's mental state during a course of psychiatric treatment is to include some elements or constructs in the grid as controls. They should fall within its range of convenience but should remain unaffected by the treatment. In a case of this kind the Queen, the Prime Minister and the Church of England were included among the elements. After a year's treatment which was unfortunately ineffectual the only noticeable change was a dramatic one in the placement of the Prime Minister. The experimenter was baffled until reminded that Mr. Heath and Mr. Wilson had exchanged roles in the interim. A control element had behaved experimentally while the experimental ones had acted as controls.

Mathematical analysis is not suitable for separating the contents of a grid into parts that must be *psychologically* significant and others that cannot be. Its proper function is to lay the contents out in an orderly way that may assist the experimenter in *his* proper task of interpreting them.

Observations inconsistent with the null hypothesis practically never occur singly. Experimental grids are sometimes simple, sometimes complicated systems of construct—element interactions. If the major component is exceptionally large there will probably also be some exceptionally large correlations between the constructs, and their entire distribution may well be far off centre. The dispersion of the elements may be dominated by a particularly salient one far distant from the rest; if so, it will generate a whole set of large interelement distances. Another possibility is that some of the elements may cluster closely; then they will tend to lie at about the same distances from ones outside the cluster. In short, the conspicuous phenomena are generally likely to be interconnected and some will help to explain others.

The null hypothesis is more likely to apply to a grid of differential changes formed for comparing two grids with the same elements and constructs, e.g. a pair of grids obtained from the same informant on two occasions, or one informant's grid and the consensus of a comparable group. Then it may be worthwhile to verify that the hypotheiss is untenable before going on to speculate on possible explanations. Quasis formed directly from real numbers should be used for this purpose.

11

METHODS FOR COMPARING GRIDS

11.1 Comparisons in general

A comprehensive review of methods which have been or could be used for comparing grids would be far too large a work to go into one chapter in a book concerned with other subjects. This chapter is mainly concerned with explaining the rationale of some programs specially written for making comparisons.

The simplest form of comparison in practice is between objects that purport to be identical but may actually differ. Thus bank notes may be compared, or postage stamps, or the cups and saucers in a china tea-set. In fact, quality control is maintained in most manufacturing processes to ensure uniformity in the products.

Much more complicated processes of comparison may be attempted between states before and after an event; for instance, when a decision has been put into effect on a large scale with some desirable end in view. Thus slums of back-to-back terraces have been demolished to make space for tower blocks in order to provide more accommodation and meet higher sanitary requirements. When the primary objectives of the programme have been achieved and the blocks built and occupied, it may be found that other social changes have been brought about, some of which may be considered regrettable. And meanwhile the way of life of the local population may have been affected by other unconnected occurrences. In comparing the states before and after such an event its effects need to be assessed in their totality, including the side effects as well as the main effects, together with its opportuneness in relation to the concomitant occurrences.

Scientific comparisons generally lie between these two extremes. They concern some deliberate alteration in a situation which is controlled as far as possible in other respects so that the effects of the alteration can be isolated. In physical experiments physical controls are applied; e.g. wind tunnels where the speed and direction of the air flow can be controlled are used in experiments in aerodynamics for measuring variations in turbulence caused by alterations in an aerofoil. In medical and psychological comparisons statistical controls are usually needed; thus in drug trials sufficient numbers of patients in matched samples are given active and inert capsules to observe the variations in their responses to their treatments and test whether the effects of the drug are significantly different from the placebo effects. In experiments of all kinds extraneous variation must be watched in case it cannot be completely controlled.

For comparisons in sporting events a suitable setting is chosen, judges are appointed and precautions of all kinds are taken to ensure that the conditions are fair to all the competitors. Rules are also prescribed for judging their relative merits.

In general it seems that for comparisons between objects or events to be precise the differences between them need to be observed under controlled conditions and assessed in accordance with specific criteria; uncertainties about either the conditions or the criteria are liable to impair confidence in the conclusions.

While it is essential not to overlook any of the conditions necessary to make a comparison impartial, it is also desirable not to impose more conditions than necessary. If a method of comparison proves reasonable the conditions attached to it may be examined to see how far they can be relaxed to extend its range of application legitimately.

Controlling the conditions for comparing grids is complicated by the extreme flexibility of the interviewing technique and the subjective nature of the data collected with it. Differences in form and content need to be considered. The formal differences can be simulated with quasis, and Chapter 10 indicates how far they extend. The sources which contribute to the contents of a grid produce far greater diversity. They include the informant, the occasion, the topic and the actual choice of elements and constructs; moreover, the interviewer's role can hardly be completely passive. Experimental designs which can isolate the contribution from each such source and measure the effects produced by their interactions have not yet been devised. At present, efforts to achieve comparability are liable to involve restricting the flexibility of the procedure or introducing questionable assumptions about the contents of the grids, or both.

Criteria proposed for comparing grids may be intended to apply to all and sundry, or only to apply under certain conditions, e.g. when the elements are persons. They may refer directly to the formal properties of grids or to their contents or they may not refer directly to either. Instead, they may be derived from a psychological theory — possibly but not necessarily personal construct theory — which leads to an expectation that grids of certain kinds will differ in certain respects. If a criterion refers to the grids directly an operational definition may be given for it which is unequivocal, but if it is derived from a theory the question may arise whether the operational definition offered for it is appropriate. The explanation for an unsuccessful experiment may be either that the theory does not apply or that the definition adopted is unsuitable.

These are some of the considerations involved in comparing grids.

11.2 The COIN program

There is, of course, no upper limit to the number of alternative indices for comparing people that can be defined by formulae applicable to data contained in their grids. Quite a wide variety of measures were reported by Bannister and Mair in 1968; the list could be extended a great deal further now and no doubt would soon need to be updated again.

Measures of intensity and consistency were proposed by Bannister (1960, 1962) after investigating evidence of thought disorder in grids of schizophrenic patients, and he published a grid test of thought disorder in collaboration with Fay Fransella in 1966. Here, the measures are calculated from two administrations of a supplied grid with eight elements which have to be ranked in terms of six constructs. These restrictions deprive grid technique of all its flexibility and convert it into a controlled testing procedure.

Bannister's definition of consistency does not necessitate the use of the same elements on both occasions, and in fact he used different, matched sets of elements in his first version of the test. The interviewer was instructed to elicit names of thirty-six people from the informant and then sort them at random into two sets of eighteen to serve as the elements in the two administrations.

COIN adopts this definition of consistency and generalizes it, extending it to examine the consistency of any number of grids. The scales for evaluating the elements may be varied. The constructs are presumed to be the same in all of them, and must be aligned — to be precise, the ith construct is presumed to be the same in every grid for all i from 1 to n. The extent to which their dispersions tend to coincide is measured by comparing the angular distances between the constructs in the different element-spaces in which they are observed. The elements need not be the same, or if they happen to be the same they need not be put in the same order in different grids. One condition is necessary: the grids should all have the same number of elements or else should have more elements than constructs. The reason is that comparability between the dispersions of the constructs in the element-spaces may be lost if some spaces are more restricted than others.

The program calculates the correlations and the angular distances between the constructs and lists them for each grid in turn. Then the average angular distances are calculated, first in degrees and then in cosines. Next the total variation in the angular distances is analysed. Suppose there are o grids with n constructs each; there will be $n(n - 1)/2$ or, say, p pairs of constructs. The average angular distance for any pair will be calculated from o values, and the variation about the average will therefore have $o - 1$ d.f. The total variation, with $op - 1$ d.f., is analysed into the total between the averages, which has $p - 1$ d.f., and the total about the averages, which has $p(o - 1)$ d.f. Finally the intraclass correlation between the angular distances is calculated and listed as a coefficient of convergence c.

The fundamental proposition is that two dispersions of n points, not necessarily in the same space, can be made to coincide if the points in each lie at unit distance from their origin, and the angular distances between the points in one dispersion are equal to the angular distances between the corresponding points in the other. Exact coincidence is impossible if one dispersion is confined to a more limited space than the other. To prove that the dispersions coincide it is sufficient to show that the distances between the corresponding points are the same; there is no need to define the operations (rotation, etc.) that have to be performed to make one dispersion coincide with the other. Similarly, if they cannot be made to coincide exactly, how closely they can be made to converge can be measured by comparing the angular distances between the corresponding points; again, the operations required to maximize convergence do not have to be defined. No further steps are needed to extend the argument to more than two dispersions.

The coefficient, c, is intended for use as a psychological measurement in the same way as a score on a test. Each of the subjects in an experiment could be asked to complete two grids with suitable specifications and the values of c could be calculated, after which groups of subjects could be compared in terms of the means and standard deviations of their measurements; or c could be correlated with other variables, etc. It is just like a consistency score in this respect, its main advantage being that it is adaptable to a wider variety of specifications. A correlation of 0.94 was found between the two measurements in a sample of forty-six cases (see Slater, 1972, where the whole subject is discussed more thoroughly).

11.3 Measurements of intensity, simplicity/complexity, and explanation power

Bannister's operational definition of intensity implies that it must depend on the proportion of the total variation in a grid attributable to its first component, or, as Jane Chetwynd (1974) says, on the explanation power of that component. In a sample of fifty cases given the Bannister—Fransella test she found a correlation of 0.98 between the two measurements.

The explanation power of the first component must vary to a large extent, depending solely on the number of constructs and elements in the grid. Compare, for instance, the ranges of values given by GRANNY for arrays of two different sizes:

One hundred arrays with the specifications	Upper Limit	Mean	Lower Limit
6 by 10 seven-point normalized	57,05	37.77	27.19
21 by 25 nine-point normalized	19.25	15.08	11.92

Of course, lower values are to be expected from quasis than from grids where there is any coherence among the constructs; this, as already mentioned, is the variable which distinguishes between experimental grids and quasis most clearly. What the figures show is that the level of explanation power which distinguishes an experimental grid from a quasi of the same size decreases as the size increases.

There are only a few elements and constructs in Bannister and Fransella's grid and they have not been chosen to allow much scope for subtle distinctions; if the first component obtained from the data does not account for a high proportion of the total variation, the explanation to be preferred may be that the thinking is disordered and not sophisticated. In a grid with a large variety of elicited elements and constructs variation may more reasonably be expected to extend widely into several dimensions. If the explanatory power of the first component is high, suspicions may arise that the informant's construct system in unduly restricted; pejorative terms like 'one-track' or 'narrow' might be applied to it. Systems where variation extends freely into several dimensions may stimulate more interest in the observer.

The construct 'cognitive complexity/simplicity' was introduced by Bieri (1955) to evaluate construct systems in this respect. The distinction is made in terms of personal construct theory, e.g. 'Presumably the more complex judge has available a greater repertory of construct dimensions along which to construe others than does the less complex judge.' The operational definition proposed for it (Bieri and others, 1966) is in terms of a 10 by 10 grid where the elements are elicited from the informant in accordance with Kelly's Rep Test procedure, i.e. as names of people who fit supplied roles; they are evaluated on a six-point scale in terms of constructs which are supplied.

The procedure given for scoring complixity/simplicity implies that the grids with the lowest scores, i.e. the simplest, are the ones where the first component accounts for the highest proportion of the variation. So the equation

intensity = explanation power = simplicity

must be true approximately. But the grids with the highest scores are actually the ones that approximate most closely to quasis. In terms of the scale the bipolar opposite of simplicity is disorganization or incoherence; where complexity belongs on the continuum is problematical.

Many other theoretical definitions have been proposed for complexity and many other methods tried for measuring it. After a comprehensive review of work done during the last twenty years, Jane Chetwynd (1974) reached the rather discouraging conclusion: 'Much more research will have to be done in both these areas before any definite criteria for the measurement of cognitive complexity can be evolved.'

The question has arisen (Crockett, 1965) whether complexity characterizes an individual as an integral personality or only his construct system concerning a particular topic. Most of us specialize in some subjects and know nothing or next to nothing about others. An orchestra conductor may have a simpler system for evaluating varieties of ice-cream than an eight-year-old. And please do not ask me to evaluate the merits of football players or fishing baits!

An adequate, operationally defined criterion is a prerequisite for investigating any such problem experimentally. If grids are compared in terms of the explanation power of the first component, there is no need to stipulate that the constructs and/or the elements they contain should be the same. The explanatory power of the component is given by INGRID for every grid as the first entry in the table listing the latent roots as percentages of the total variation; and grids may be compared in terms of it provided they are of the same size, n by m. Thus a 'things' grid and a 'persons' grid of the same size but with different elements and constructs could be elicited from an informant and compared to see in which the explanatory power of the first component is higher. Whether it should be called the simpler grid, the more coherent one or the one with the greater intensity would be a question of interpretation.

11.4 What is the same as what?

Formal criteria, i.e. criteria like the explanation power of the first component, that can be applied to grids matched only in some formal respect such as their size, have certain advantages for purposes of comparison. The conditions under which they apply are very general and can be defined without much risk of being misunderstood by anyone. No wonder they have received a great deal of attention.

But the more widely the definition of grids is extended the less comparable the psychological implications of any of their formal properties are likely to become. Comparisons referring to the contents of particular grids that are appropriately and closely matched are more likely to provide evidence relevant to particular psychological problems.

Unfortunately, awkward questions arise when any attempt is made to match two grids in terms of their contents, particularly when the grids have been obtained from different informants. Do the same verbal labels, attached to the elements or the constructs, convey the same meaning to both of them? Suppose, for example, that both use nice/nasty, pretty/ugly and kind/unkind as constructs to evaluate an agreed set of mutual acquaintances. When their grids are analysed, nice/nasty may be found to be almost synonymous with kind/unkind in one, but not with pretty/ugly, while in the other grid the relationships may be reversed. Is this because the constructs do not have the same meaning for both the informants? Or because they have assessed their acquaintances from different points of view? Or because the interactions between the constructs and the elements are affected by special relationships between some of the parties concerned?

Such questions are not evaded when the grids have been obtained from the same informant on different occasions. If he does not evaluate the elements in the same way in terms of the constructs on each occasion all three alternative explanations remain open for consideration. The changes may be attributable to some of the constructs or some of the elements or some of their interactions; and these alternatives are not mutually exclusive.

The devout adherent to personal construct theory may parry such questions by quoting from Kelly's (1955) corollaries (particularly the commonality and sociality corollaries) to prove that he is not deviating from the theory when he makes assumptions that functions in one grid can be equated with functions in another. The computer programmer can evade them too. He need only say that such questions must be answered by the experimenter before submitting his data for analysis.

If a sceptical philosopher mindful of Heraclitus were asked the plain question, 'Is this function the same as that?', he could hardly be expected to give the plain answer, 'Yes'. The question should be put to him as 'Is it reasonable to treat this function as if it were the same as that?'. The general purport of his answer might then be that it seems more reasonable in some circumstances than others. For instance, if viewers of a television programme are asked whether they found the items in it exciting, amusing, informative, boring, etc., or not, the differences in their opinions may reasonably be regarded as attributable to differences either in the interpretations of the constructs or in the construct/element interactions. The third alternative may be ruled out.

Or suppose names are elicited from different informants to fit supplied roles in accordance with Kelly's Rep Test procedure. Different men, no doubt, will be found filling the role of father in different grids. But the experimenter may feel entitled to combine them all as members of the same class, say for the purpose of finding out what view of 'father' is typical of children in a particular community, or whether one child's view of his own father is different from the typical one.

The lexicographer may help. He will point out the importance of distinguishing between the connotation and the denotation of words. People do not disagree about what the word 'edible' means, for instance, just because they disagree about how some of the articles offered for human consumption should be construed.

An experimenter should perhaps feel fortunate if the final decision is left to him. Yet some may not; they may prefer to have their decisions forced on them and to convince themselves that their conclusions are inescapable. It may be distressing for them to discover that the conclusions they will reach are dependent on the answers they choose to give the question, 'What is the same as what?'. But so they are.

11.5 A general plan for comparing grids

The number of ways in which grids can be compared is evidently limitless. To make an orderly approach to the subject, let us begin by considering ways of assembling a set for comparison, supposing that they have been collected from one informant by asking him to complete a grid with the same m elements and n constructs on o occasions. Such sets of grids can be obtained from patients during a course of psychotherapy, and have in fact been collected and studied, for instance by Feldman (1975).

The letter i is taken to indicate any construct in the range from 1 to n, j for any element from 1 to m, and k for any occasion in the series from 1 to o. The whole collection can be assembled as a three-way tensor defined mathematically as

$$\mathbf{T} = [t_{ijk}]$$

t_{ijk} being its characteristic entry. It is like a book with o pages, where the data obtained on the first occasion are given on the first page in the form of a table with n rows and m columns, the data from the second occasion on the next page in the same format, and so on. The contents of the first page may be denoted $\mathbf{T}_{..1}$, i.e. all the values of t_{ijk} for $k = 1$, and successive arrays denoted likewise through to $\mathbf{T}_{..o}$, the entries on the last page.

In some ways it is more like a pile of bricks, for it is just as easily separated by row or by column as by level. For instance, all the entries referring to the construct i, which appear in the ith row on every level, can be extracted to form a two-way array, $\mathbf{T}_{i..}$, and all the entries for element j likewise form the two-way array $\mathbf{T}_{.j.}$.

If the experimenter feels reasonably certain that the same constructs have been used in all the grids but knows or suspects that some of the elements have changed, he may prefer to set the grids out side by side to form a single two-way array with n rows and om columns. Then each element will be represented geometrically by o points in the n-dimensional construct-space; they may be regarded as marking out the path along which the element has moved from one occasion to the next. For example, in a study of transference, changes in a patient's attitude towards his therapist could be traced if he completed a grid from time to time during the course of therapy, including his therapist and himself as elements.

Or if the experimenter can assume that the grids all refer to the same elements but suspects that the constructs have been used differently, he may decide to string the grids together one beneath another to form a single two-way array with no rows and m columns. This would be appropriate if the series consists of different informants' evaluations of the same elements instead of one informant's evaluations of them on different occasions. For instance, if the experimenter wants to find out how o observers apply the construct 'interesting' to the items on a television programme, he may assemble the data in this way to see how the different uses of the construct are distributed in the element-space. If the ratings do not agree we may say that the observers' constructs are effectively different, though nominally the same.

These three ways of assembling the data correspond to the three alternatives mentioned in the last section for explaining the variation they record. Assembling the grids into a two-way array extended lengthwise holds the element-space stationary and examines differences in the dispersion of the constructs; thus it leads to a complete analysis of the observed variation in terms of differences between the constructs. Assembling them into a two-way array extended sideways holds the construct-space stationary and accounts for the total variation in terms of differences between the elements. Accumulating them as a three-way array holds both sets of reference axes stable and accounts for the variation in terms of the interactions between them.

Each procedure leads to a complete analysis of the data. Presumably the experimenter will choose the one that appears to fit his expectations best. But he should not ignore the other alternatives in case one turns out to be unexpectedly illuminating.

Grids which are matched by element but not by construct can be assembled lengthwise, but not in either of the other ways; the constructs need not even be the same in number. Similarly, grids matched only by construct can be assembled sideways, but the other alternatives are excluded.

The alternatives are illustrated diagrammatically in Figure 11.1. Grids are shown spread out horizontally and vertically and piled up perpendicularly, depending on whether they are matched by construct, by element, or by both. Grids not matched in either respect can be included, spread out diagonally. Grids with some elements or constructs in common, or both, could be visualized as adjoining or overlapping to some extent. The implication is that the closer the contact between the grids the more scope there is for comparing them. Grids with no elements or constructs in common can be compared in terms of their formal properties (see also Section 11.11).

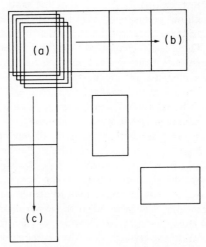

FIGURE 11.1 Alternative ways of assembling grids for analysis. Grids with the same constructs and elements can be assembled in a three-way array (a), or in a two-way array by construct (b) or by element (c). Grids with the same constructs but different elements can be assembled horizontally (b) but not otherwise; grids with only their elements in common can only be assembled vertically (c). There is no way of assembling grids without constructs or elements in common

11.6 PREFAN

There are no mathematical problems involved in adapting principal component analysis to the two-way arrays formed by assembling grids horizontally or vertically beyond those involved in the analysis of a single grid by INGRID. The operations are performed in a different way to accommodate the extra data, and some parts of the results have to be omitted as the output might otherwise become overwhelming.

The program for analysing a vertical assembly of grids, PREFAN, is a modification of an earlier one written to carry out a method of preference analysis (Slater, 1960). The data were to be obtained by asking informants to evaluate a given set of items according to their own personal preferences. The analysis then converges on Q-technique, as it amounts to a principal component analysis of the correlation between the informants. The program still retains vestiges of its origin in its present form, the elements being described in the output as items and the constructs as cases; and it can still be used as originally intended.

In order to save trouble for an experimenter who may want to have his grids analysed both individually by INGRID and collectively by PREFAN the same format is used for punching them for both programs. The data are stored in an n by m array with the limits $n \leqslant 500$ and $m \leqslant 40$. The constructs are centered and standardized (converting V_i to $m - 1$ for all i), and the resulting array of deviations, $_n\mathbf{D}_m$, is premultiplied by its transpose to form an m by m product matrix, to which the analysis is applied.

The particulars of the grids accepted are copied from their pilot cards and listed at the beginning of the output. The sums of the entries in $\mathbf{D}$ referring to each item are given next, together with their sums of squares, expressed quantitatively and as percentages of the total variation ($V = S = n(m - 1)$). Then the product matrix $\mathbf{D}'\mathbf{D}$ is listed and followed by the table of standardized distances between the items. (For explanations and examples see Section 7.9 and 9.2 and Tables 9.3 and 9.4. The product matrix $\mathbf{D}'\mathbf{D}$ is also denoted $\mathbf{P}$, as in Section 8.2.)

The output concludes with the results from the principal component analysis: the latent roots of the components, expressed as actual amounts and as percentages of the total observed variation; the results from the Bartlett test if required; and the full specifications for as many components as required. The vectors, loading and polar coordinates from the first three components are listed for the items, and so are the loadings and coordinates for the constructs in each case. Lastly, the item vectors and the construct loadings per case are listed for any subsequent components included in the options or indicated by the Bartlett test (see Section 8.2 to 8.7, 9.3 to 9.6, and Tables 9.5 and 9.6). Experience suggests that the assumptions of the Bartlett test are more appropriate when several grids are assembled for analysis by this program than when they are analysed individually by INGRID.

The results generally found most interesting are the ones for the first three components. They can be exhibited graphically as in Figure 9.3 or, better still, displayed on the surface of a globe as explained in Section 8.7 (see also Section 9.5). When the number of constructs is large enough to warrant using PREFAN in place of INGRID, listing the correlations between them is not practical either for the computer or the experimenter. One additional construct, increasing their number from n to $n + 1$, adds n to the number of correlations between them:

it is over 1,000 when n exceeds 45 and goes on increasing at a constantly accelerating rate with each addition to n. On the other hand, each addition to n only adds one point to the three-dimensional plot, and the angular distances between it and the other points are easy to see. Some oversimplification is unavoidable — how much depends on how far the total variation extends beyond the space of the first three components.

11.7 ADELA

ADELA analyses a set of grids where the constructs are regarded as the same but the elements are free to vary; the construct-space is held stable and the additional elements are located in it according to their coordinates. Like PREFAN it can accept any grids acceptable by INGRID, with the same pilot cards and variable format cards. They are stored in an n by m array with n up to 40 and m up to 500 (there is also a modified version for arrays up to 20 by 1,500). Centering by construct may be applied to each grid separately or to the array as a whole, and normalization may be applied to each grid separately or to the array as a whole, or not applied at all. There is no separate provision for ranked grids — all are treated as graded.

The program can also accept a single n by m array of data, reading it in either by row or by column. When the data are assembled in this way the analysis converges on R-technique. The array is formally indistinguishable from an array of the scores of m subjects on n tests, and the results obtained after normalization are equivalent to a principal component analysis of the correlations between the tests. It is applied to the n by n product matrix formed by postmultiplying the array of deviations by its transpose.

When the data all come from one grid the output begins with a list of the mean and total variation of each construct and the total variation for all constructs combined. If the constructs are not to be normalized their covariance matrix is listed next and followed by the correlations and the angular distances between them, but if normalization is applied the covariance matrix is omitted. The specifications for the elements come next, viz. their sums and sums of squares, the latter expressed both in the metric employed and as percentages of the total variation. The program then goes onto the principal component analysis.

When several grids are assembled their specifications are copied from their pilot cards and listed. When all the data have been read in they may be treated immediately as a single array; if so, the output from it will take the form already described. Otherwise, each grid will be centered separately, and an option is available to obtain a list of the mean and total variation of each construct in each grid. The grids may also be normalized separately; if so, the output continues with the table of correlations and angular distances between the constructs calculated from the whole array of data. If not, the covariance matrix between them is inserted first. Thus data subjected to different optional processes arrive at the principal component analysis by different routes.

Measurement of the distances between the elements are omitted from ADELA because they would be unmanageably bulky, like the correlations between the constructs that would be obtained from PREFAN.

The specifications of the principal components are given in the same way as by

INGRID and PREFAN. The latent roots are listed, with the results from the Bartlett test if required. If so, the experimenter may set upper and lower limits to the number of components to be specified in full; if not, he can obtain specifications for as many as he wants by setting the same value for both limits. The vectors and residuals given are for the constructs in the array as a whole and for each element in each grid. Polar coordinates are listed for the first three components if full specifications for three or more are required. Construct vectors can also be obtained for minor components beyond the number to be specified in full.

11.8 Three-way arrays

One might hope that a residual three-way array of deviations could be derived from T by some preliminary operations comparable to centering and then be referred to components with properties like those of two-way arrays of deviations. They would be strictly orthogonal and their latent roots would form an orderly series accounting for successively smaller amounts of the total observed variation, the only difference being that the specifications of each component would include three vectors, namely a serial vector as well as a row vector and a column vector.

This would provide an attractive way of interpreting the data. For instance, grids obtained from o observers might be referred to a common set of components, such as Osgood's (1957) dimensions of evaluation, activity and potency, the variation between the grids being attributable to differences in the values assigned to the elements and the importance attached to the dimensions by the observers. Unfortunately such an explanation must generally force an oversimplification onto the data.

John Gower's comprehensive discussion of the whole problem (see Chapter 12) shows that such a simple solution cannot be reached as a rule. Two-vector product terms need to be calculated as well as the three-vector terms, and they are not independent (see equation 12.8). The general problem of making a canonical analysis of a three-way tensor is not yet solved and is likely to remain unsolved for some time yet, in spite of the many attempts that have been made to reach a solution. Nor can we feel certain that the optimal solution, when it is reached, will be defined in mathematical terms that will be convenient for psychological interpretation. The methods described here for analysing three-way arrays are additive, not multiplicative.

11.9 DELTA

Comparing grids aligned by construct and element is simplest when there are only two to be considered. We may suppose for convenience that they are obtained from the same informant on two occasions and that the question for consideration is whether they provide evidence of any change in his state of mind. These are, of course, not the only circumstances in which two completely aligned grids may need to be compared, and in other circumstances the question for consideration may be defined differently. The mathematical procedure still remains the same. The essential operations are to form a grid of differential changes by subtracting the first grid from the second, and then to put it through a principal component analysis. The results show the extent and direction of the changes that have occurred.

The suffix (a) is used to identify the terms which characterize the first grid, and (b) for the second. Thus in reference to row i of $G(a)$ the mean of the entries is denoted $g_i.(a)$; their deviations from the mean form the vector $d_i.(a)$; and the sum of the squares of these deviations is $V_i(a)$ — see Section 5.2. The corresponding terms for $G(b)$ are $g_i.(b)$, $d_i.(b)$ and $V_i(b)$.

The two arrays of deviations from the construct means, $D(a)$ and $D(b)$, are not normalized, since the psychological implications of the changes in the variation recorded by a particular construct on different occasions cannot safely be ignored. Distinctions between the elements in terms of it may seem important at one time and trifling at another. For those who feel a need for statistical reassurance the ratio $V_i(a) : V_i(b)$ can be entered as a variance ratio with $n_1 = n_2 = m - 1$ as a test of significance.

The grid of differential changes, Δ, is formed as

$$\Delta = [\delta_{ij}] = [d_{ij}(b) - d_{ij}(a)]$$

That is to say, the value of δ_{ij} measures the relative change in the evaluation of element j compared with the others on construct i from the first occasion to the second. If its change does not differ from the mean change, $\delta_{ij} = 0$.

The covariation between the sets of entries for construct i on the two occasions is

$$d_i.(a) \cdot d_i'.(b) \quad \text{or, say} \quad w_{ii}$$

and the correlation between them is

$$\frac{w_{ii}}{(V_i(a) \cdot V_i(b))^{\frac{1}{2}}} \quad \text{or, say,} \quad r_{ii}$$

If we wish to know whether the average rating of the elements on construct i was significantly higher on one occasion than the other, we may apply a correlated t test. The standard error (s.e.) of the difference is obtainable from the formula

$$\text{s.e.}^2 = \frac{V_i(a) + V_i(b) - 2w_{ii}}{m(m-1)}$$

The enumerator in this equation is equal to

$$\delta_i. \cdot \delta_i'. \quad \text{or, say,} \quad V_i(\delta)$$

i.e. the sum of the squares of the entries in row i of Δ.

The decks of cards for the two grids to be compared should each be punched in a format suitable for analysis by INGRID and they should be read in with $G(a)$ first. After copying the identifying particulars for the two grids from their pilot cards the program lists the important quantities derived in forming Δ, under the general heading

CONSTRUCT MEANS, TOTAL VARIATION, etc.

The values listed are $g_i.(a)$ and $V_i(a)$ under the subheadings *mean* and *variation* for Grid A; $g_i.(b)$ and $V_i(b)$ similarly for Grid B; the difference, $g_i.(b) - g_i.(a)$, and the variation, $V_i(\delta)$, under the headings *mean* and *variation* of the *changes $B - A$*; then the *correlation*, r_{ii}, with its s.e. and the value of t. After these quantities have all been listed for constructs $i = 1, \ldots, n$, the cumulative totals $V(a)$, $V(b)$ and $V(\delta)$ are given and, finally, the *general degree of correlation* between the two grids. It is calculated as

$$\frac{\Sigma w}{(V(a) \cdot V(b))^{\frac{1}{2}}}$$

where $\Sigma w = w_{11} + \ldots + w_{ii} + \ldots + w_{nn}$.

When the grids are ranked the process of forming Δ is considerably simpler. It reduces to

$$\Delta = \mathbf{D}(a) - \mathbf{D}(b)$$

(to assign a positive value to a rise in rank from the first occasion to the second).

The only quantities that need to be specified for each construct are $V_i(\delta)$ and r_{ii}. They are listed as the *total variation* and the *correlation per construct* under the general heading

EFFECTS OF CHANGES IN RANK

Finally, the value of $V(\delta)$ is listed as the total for *all constructs*.

The value of V_i, which is the same for every construct in each grid, is included as the *total variation per construct per grid*; then the total for all the constructs, $nV_i = V$, is given as the *total per grid*. This is followed by the *general degree of correlation*.

The grid of differential changes, Δ, has the same form whether the grids it is derived from are ranked or graded. It is printed out under the heading

GRID OF DIFFERENTIAL CHANGES, UPRATINGS POSITIVE

If an element is rated (ranked or graded) differentially higher in B than A its entry is positive for the construct concerned.

The analysis applied to Δ is an abbreviated version of the analysis used for an individual grid in INGRID 72, except that no option to normalize is provided and the program terminates without listing the relations between constructs and elements or the relations of the elements with one another in terms of angular distances and cosines. Should results of this kind be required an option is available to instruct the computer to punch the grid of differential changes onto a deck of cards which can be accepted for complete analysis by the INGRID 72 program. The sequence in which the results from DELTA are printed out follows the sequence described for INGRID 72 in Chapters 7 and 8. The comments which follow point out the most important differences found in the psychological contents of the results when the computing program is applied to a grid of differential changes.

In the table of correlations and angular distances between constructs the correlation between construct 1 and 2, for example, is the correlation between changes in the application of construct 1 to the elements and the changes in construct 2. A high correlation indicates allied or opposed changes in the application of the two constructs, according to sign. A small correlation is most likely to occur when there has been little change in the application of one or another, or both, but may also occur when the constructs have changed independently. The choice between these alternatives can be decided by examining the relevant entries in Δ.

In the results listed under heading

ELEMENT   TOTAL   SUM OF SQUARES

the elements with the greatest sums of squares (value of $S_j(\delta)$) are the ones which show the largest changes. The list of sums of squares is in fact a breakdown of $V(\delta)$ $(= S(\delta))$ by element. A large entry in the column TOTAL for an element indicates, according to sign, that it tends to be uprated or downrated consistently in terms of all the constructs; a small entry may indicate no large changes or changes in opposite directions for different constructs.

In the table of distances between elements a large entry indicates divergent changes in the evaluation of the two elements concerned. A small one may indicate changes in the same direction, or little or no change in the evaluation of either. As before, the decision between these alternatives can be reached by referring to Δ.

The list of the roots of the components is a breakdown of $V(\delta)$ into the amounts attributable to the successive components. The construct loadings of the major component show how the major shift from **A** to **B** is related to the constructs, and its element loadings show how the shift is related to the elements. It may sometimes be found to be related to certain elements in particular and spread widely over the constructs, or it may be related particularly to certain constructs, affecting most of the elements to some extent. It is advisable, in each case, to consider the changes which are particularly conspicuous first and then proceed to the more widespread ones. Examining only how the shift affects the constructs, and overlooking the evidence for it among the elements, may not involve missing much in every case, but it will in some cases. Minor independent shifts in other directions will be defined by the specifications of the minor components.

If there are only a few slight differences between $G(a)$ and $G(b)$, the values of r_{ii} and the general degree of correlation between them will tend to be uniformly high and the values of $V_i(\delta)$ and $S_j(\delta)$ uniformly small. In short, there will be several points in the evidence to indicate that the construct system has remained stable.

Instability may appear in many different ways. It may be of a diffuse, unsystematic kind, indicating that the informant is uncertain how the constructs should be applied to the elements; or it may be of a specific kind, indicating uncertainty about how a particular construct applies to the elements or how a particular element should be evaluated in terms of the constructs. In that case one exceptionally small value of r_{ii} or exceptionally large value of $V_i(\delta)$ or $S_j(\delta)$ will appear among the results. Occasionally the difference between the two grids may be located even more specifically in the interaction between a particular element and a particular construct; then it will appear as a particularly large entry in Δ.

If intervening events have produced a coherent reorganization of the informant's construct system, the principal component analysis of Δ should serve to indicate the extent and direction of the changes that have occurred in ways that will help to simplify interpretation. In case of doubt the results from the analysis should be compared with results obtained from GRANNY, using quasis of non-normalized real numbers with the same n and m.

One possible use of DELTA is to measure effects of psychiatric treatment. The treatment may be directed towards modifying either the way in which certain elements are construed or the way in which certain constructs are applied. An obsessional neurosis, for example, may be characterized by abnormal evaluations of certain elements; a neurotic depression by the abnormal application of certain constructs; and treatment will be directed towards them in each case. They must be taken to be the experimental elements or constructs as the case requires, but other functions should be included in the grids besides the ones expected to be influenced directly. Some may be worth inclusion to serve as controls and some to bring possible side-effects within reach of observation. Then data will be obtained for comparing changes attributable to treatment with other changes.

11.10 SERIES and SEQUEL

When o grids

$$G(a), G(b), \ldots, G(o)$$

are collected referring to the same n constructs and m elements a reasonable way of analysing them completely is to extract a consensus grid from them, analyse it and then go on to compare each individual grid with the consensus. The consensus grid is formed by the program SERIES. Afterwards it can be analysed by INGRID and compared with the individual grids by an iterative form of the DELTA program called SEQUEL.

SERIES performs all the preliminary processes involved in forming the consensus grid, which is listed in the output as B. It is defined as

$$_n\mathbf{B}_m = \frac{\mathbf{D}(a) + \mathbf{D}(b) + \cdots + \mathbf{D}(o)}{o}$$

that is to say, it is the mean of the set of arrays formed by centering each grid by construct. It can be punched onto cards if desired, as well as listed. The output includes a good deal of extra information as well as the consensus.

In its present form SERIES accepts up to 250 grids with fifty elements and constructs. They may be graded or ranked. They should be punched in the same format as for INGRID. The title of each grid is listed as it is read in and the array of deviations formed from it is written on magnetic tape (identified in the program as tape 20). After all the data have been read in, the results for each construct are listed in turn.

The raw data referring to a particular construct may be treated as a two-way array with a row for each occasion and a column for each element, and an analysis of variance may be applied to it. The complete set of entries has a general mean, the means of the o rows vary about it and so do the means of the m columns. The residual variation may be attributed in the usual way to the element/occasion interactions, but can be partitioned further into separate amounts for each of the elements.

The output for the construct begins with the general mean. Then the means for the o occasions are listed as deviations from the general mean at (A), and the means for the elements are listed likewise at (B). The separate amounts contributed by the elements to the residual variation are given next. A small amount of information is added concerning the variation about the construct means on different occasions, namely the occasion when the variation was greatest and when it was least, and the amounts observed on those two occasions.

The output for the construct concludes with a summary of the results from the analysis of variance: the total variation due to differences between occasions (source A, with $o - 1$ d.f.); the total due to differences between elements (source B, with $m - 1$ d.f.); and the total due to occasion/element interactions, which has $(o - 1)(m - 1)$ d.f. Comparing the mean squares from source B with the residual mean square will reveal how much agreement there is on different occasions about the evaluation of the elements in terms of the construct. Comparing the amounts contributed by separate elements will reveal whether agreement is closer about some elements than others.

The lists of element means recorded at (B) for the successive constructs form the rows of the consensus grid. If an INGRID analysis is to be applied to it or if it is needed for comparisons with the individual grids by SEQUEL, the option should be taken to have it punched out on cards.

The lists of occasion means recorded at (A) for the successive constructs could be assembled into an n by o matrix which would show what interactions have occurred between these two sources of variation. It might be of interest in some cases to see, for instance, whether fluctuations occur during the course of time in the relationship between one set of feelings, such as guilt, depression or anxiety, and another set, say of confidence, friendship or dependence. But I cannot cite any experimental work along these lines. SERIES is more often applied to a set of grids with the same elements and constructs obtained from o observers than from the same informant on o occasions. Under these conditions the results from the (A) matrix are less likely to be of interest.

Ranking simplifies the computations and the output considerably. Every construct in every grid has the same mean and variation; these two quantities only need to be recorded once, immediately after the grids have been read in and listed. Next comes the total variation for each construct on all occasions combined, i.e. the total concerned in each analysis of variance. The output for each construct gives the differences between the element means and the general mean, which constitute the entries in the appropriate row of the consensus grid, and then the amount contributed by each element to the residual variation. For the concluding analysis of variance the only results needed are the total variation due to differences between the element means, with $m - 1$ d.f., and the total due to element/occasion interactions, with $(m - 1)(o - 1)$ d.f. Ranking sacrifices $o - 1$ degrees of freedom by eliminating differences between occasions as a source of variance.

An experimenter may collect a set of doubly aligned grids from certain observers, not because he is interested in any of them personally but because he wants to speculate about possible results from experiments he intends to avoid making with other observers. How confident can he be that they will rate one element higher than another on a given construct, because the observers in his experiment did so? Or that they will rate a given element higher on one construct than another, from corresponding evidence? Statistical methods are meant to answer such questions.

For both comparisons a correlated t test is suitable. The covariance required to calculate the standard error of the difference between the means of two elements on a given construct depends on the covariance between their ratings on it.

Conversely the s.e. of the difference between the means of a given element on two constructs depends on the covariance between its ratings on them. In the first case, say the elements are a and b and the construct is x; then the covariance required is the summation of the products $d_{xa} \cdot d_{xb}$ over the o occasions. In the second case, say the element is a and the constructs are x and y; the covariance required is the summation of $d_{xa} \cdot d_{ya}$.

A subroutine, EXTRAS, is appended to SERIES to provide relevant results. For each construct it lists

(a) the covariance between the elements and
(b) the standard error of the difference between the means for each pair of elements;

and for each element it lists

(c) the covariance between the constructs and
(d) the standard error of the difference between its means on each pair of constructs.

Options can be entered on the pilot card for the data read in by SERIES to obtain any combination of these four sets of results.

SEQUEL uses the consensus grid, B, punched out by SERIES and the data listed by it on tape, viz. $D(a)$, $D(b)$, . . . ,$D(o)$. It begins by listing B and then forms a series of arrays of deviations from consensus:

$$\Delta\,(a) = D(a) - B$$

etc. to

$$\Delta\,(o) = D(o) - B$$

Each of these is analysed by the DELTA program, which is incorporated as a subroutine in SEQUEL and has already been described.

11.11 Some further notes on comparing grids

Besides having their separate uses the programs described can be combined in various ways. For example:

Suppose grids referring to the same constructs and elements are obtained from several groups of subjects. The data could be analysed by SERIES for each group separately, and again for all of them treated as a single group. Then the total variation for each construct could be analysed into the amounts between groups and the amount within. Or the consensus grids for the separate groups could be compared by DELTA. Or grids obtained from individual members of one group could be compared with the consensus from another, using SEQUEL.

Or if doubly aligned grids were obtained from a group of subjects before and after some course of treatment, SERIES could be applied to the two sets of data to obtain consensus grids; these could be compared by DELTA to see whether the changes exhibited any prevalent pattern. More might be learned, however, by applying DELTA to the two grids obtained from each subject and getting the grids of differential changes punched out for analysis by SERIES. In this way the changes revealed in the grids for each subject would be measured separately as well as the prevalent pattern of change. That might be disappointingly small, not because the treatment was ineffectual but because it affected different subjects in different ways.

As was explained in Section 11.5, a set of o grids referring to the same n constructs and m elements forms a three-way array which can be partitioned in other ways besides being taken apart in the way in which it was assembled. It can also be separated into m arrays, each n by o, or into n arrays, each m by o. Partitioning the data by element is most likely to be useful when the grids have been obtained from the same informant on o occasions. The data referring to one element, say 'myself as I am', could be extracted and analysed by INGRID to examine how the informant's evaluations of himself change over the series of occasions. Or the data referring to a pair of elements, say 'myself as I am' and 'myself as I would like to be' (or 'myself as I would like to be' and 'my therapist'), might be extracted and compared by DELTA. In this way changes in the relationship between the two self-images (or in the transference relationship) might be examined. Phenomena of this kind are presented and discussed in Chapter 5 of *Explorations* (see also Chapter 8 there).

It would also be possible to extract the data referring to two constructs to form two m by o arrays and compare them with *Delta*. The relationship between 'strong' and 'cruel', for instance, might be taken as a topic for study in a particular case. I have not come across any study of this kind, however, and am not certain how it would work out. Would it be better to centre the arrays by element or occasion? Either way, the procedure would lead to a different view of the data from any considered here, where all analyses begin with centering by construct, i.e. by operator.

In some circumstances the condition that the grids to be compared must be aligned by construct and element, as required for DELTA, SERIES and SEQUEL, may be found to be inconveniently restrictive. Even the condition that one set of functions must be held construct while the other if free to vary, as required by PREFAN and ADELA, may be somewhat artificial. More freedom may be desirable, for instance when grid technique is used as part of the treatment during a course of psychotherapy. When the therapist and the patient examine the results of one grid together they may come to the conclusion that some important terms have been omitted from it or that some of the ones in it are not particularly relevant. They may feel the need to reconstitute it by introducing or discarding elements or constructs, or exchanging their functions, in order to adapt it to the current occasion. The presenting symptoms of a neurotic disturbance often serve only as pointers to the mental processes that activate and sustain it.

Provided some functions are retained in the grids in the series, comparison may be made between them in terms of results obtained from INGRID, such as correlations, linear distances and angular distances, which are all measurements expressed on standardized scales. They are liable to vary, of course, being affected by the conditions under which they are observed, just as temperatures, wind velocities, etc., vary under different climatic conditions. But questions as to why they vary are just as reasonable to ask since they are expressed in commensurate terms.

Comparisons between the contents of grids may be made even when none of the terms they contain are matched. For this purpose some classificatory system must be adopted. The best known by far is Osgood's (1957) classification of scales under the three main headings of evaluation, activity and potency. Grids which have nothing more in common than a topic of interest, such as 'people I know personally', could be elicited from a number of subjects and compared simply in terms of the relative frequency of use of constructs of different kinds.

A much more detailed classification of constructs applicable to people as elements has been proposed by Landfield (1971). He defined twenty rating categories with satisfactory interjudge reliability. Topçu (1976) used this system of classification in comparing grids of seventy-five patients at Netherne Hospital, thirty with case histories including incidents of overt aggression, fifteen with histories of attempted suicide and thirty whose histories did not include episodes of either kind. He distinguished further between the explicit and the contrasted poles of the constructs, noting for instance whether the informant pointed out rejected elements first and then contrasted them with approved ones, or vice versa. Interesting characteristic differences were disclosed between the construct systems of the three groups of patients.

Since every aspect of human experience can serve as a source for the elements of a grid and can be described by a specially appropriate set of constructs, the endlessness of the possibilities for comparisons is almost overwhelming.

REFERENCES TO PART II

Bannister, D. (1960). 'Conceptual structure in thought-disordered schizophrenics'. *J. Ment. Sci.*, **106**, 1230–1249.

Bannister, D. (1962). 'The nature and measurement of schizophrenic thought disorder'. *J. Ment. Sci.*, **108**, 825–842.

Bannister, D., and Fransella, F. (1966). 'A grid test of schizophrenic thought disorder'. *Brit. J. soc. clin. Psychol.*, **5**, 95–102.

Bannister, D., and Mair, J. M. M. (1968). *The Evaluation of Personal Constructs*, Academic Press, London.

Bartlett, M. S. (1950). 'Tests of significance in factor analysis'. *Brit. J. stat. Psychol.*, **3**, 77–85.

Bartlett, M. S. (1951a). 'The effect of standardisation on a chi-squared approximation in factor analysis'. *Biometrika*, **38**, 337–344.

Bartlett, M. S. (1951b). 'Further note on tests of significance in factor analysis'. *Brit. J. stat. Psychol.*, **4**, 1–2.

Berg, I. A. (1967). *Response Set in Personality Assessment*, Aldine, Chicago.

Bieri, J. (1955). 'Cognitive complexity– simplicity and predictive behaviour'. *J. abnorm. soc. Psychol.*, **51**, 263–268.

Bieri, J., and others (1966). *Clinical and Social Judgement*, John Wiley and Sons, New York, London, Sydney.

Carnap, R., and others (1971). *Studies in Inductive Logic and Probability*, University of California Press.

Chetwynd, S. J. (1974). *Generalized Grid Technique and Some Associated Methodological Problems*, Ph. D. thesis, University of London.

Crockett, W. H. (1965). 'Cognitive complexity and impression formation'. In B. A. Maher (Ed.), *Progress in Experimental Personality Research*, Vol. 2, Academic Press, New York.

Feldman, M. M. (1975). 'The body image and object relations'. *Brit. J. med. Psychol.*, **48**, 317–332.

Hotelling, H. (1933). 'Analysis of a complex of statistical variables into principal components'. *J. educ. Psychol.*, **24**, 417–441 and 498–520.

Kelly, G. A. (1955). *The Psychology of Personal Constructs*, Norton, New York.

Keynes, J. M. (1921). *A Treatise on Probability*, Macmillan, London.

Landfield, A. (1971). *Personal Construct Systems in Psychotherapy*, Rand McNally and Co., Chicago.

Osgood, C. E., Suci, G. J., and Tannenbaum, P. H. (1957). *The Measurement of Meaning*, University of Illinois Press, Urbana, Chicago and London.

Pearson, E. S., and Hartley, H. O. (1958). *Biometrika Tables for Statisticians*, Cambridge University Press.

Slater, P. (1951). 'The transformation of a matrix of negative correlations'. *Brit. J. stat. Psychol.*, **6**, 101–106.

Slater, P. (1960). 'The analysis of personal preferences'. *Brit. J. stat. Psychol.*, **13**, 119–135.

Slater, P. (1972). 'The measurement of consistency in repertory grids'. *Brit. J. Psychiat.*, 121, 45—51.

Topçu, S. (1976). *Psychological Concomitants of Aggressive Feelings and Behaviour*, Ph. D. thesis, University of London.

Watson, J. P. (1970). 'The relationship between a self-mutilating patient and her doctor'. *Psychother. Psychosom.*, 18, 67—73.

Wilkinson, J. H. (1965). *The Algebraic Eigenvalue Problem*, Clarendon Press, Oxford.

PART III

METHODOLOGICAL CONTRIBUTIONS

12

THE ANALYSIS OF THREE-WAY GRIDS

J . C. Gower

Rothamsted Experimental Station

12.1 Introduction

This chapter is concerned with the combined analysis of more than one grid. A generalization of the methods used for single grids is presented below, but it should not be forgotten that in recent years other methods have been developed for analysing similar sets of two-way tables that may well give useful additional information when applied to sets of grids. Sets of grids are of two kinds: (a) those collected for one person on different occasions and (b) those collected for several persons but with the same constructs and elements. In the following no distinction is made between these different situations and it is to be hoped that methods will soon be developed that avail themselves of any information there may be in the time series nature of type (a) data. The elements may be ranked or any commensurate real numbers can replace the rank scores.

There are three main types of analysis other than the one to be discussed in detail below. These are individual scaling, three-dimensional extensions of factor analysis and methods based on analysing matrices whose values are some symmetric function measuring the (dis)similarity of pairs of grids. A key reference for individual scaling is Carroll and Chang (1970), who seek to represent the ith individual by coordinates given in matrix $\mathbf{X}^{(i)}$, which can be expressed in the form $\mathbf{X}\mathbf{D}^{(i)}$, where $\mathbf{X}$ represents average coordinates for the whole group and $\mathbf{D}^{(i)}$ is a diagonal matrix of weights that the ith individual gives to the group axes. The computational process involves what Carroll and Chang (1970) term the general canonical decomposition of a multiway array $\mathbf{Z}$ in the form $z_{ijk...} = \Sigma_{t=1}^{r} a_{it} b_{jt} c_{kt} \ldots$, where r is specified. Clearly this may be regarded as a generalization of the singular-value decomposition of a matrix into a product of orthogonal matrices (often known to psychologists as Eckart–Young decomposition, following Eckart and Young, 1936). The three-way canonical decomposition of a table is closely related to the work described below and, as Carroll and Chang (1970) pointed out, has links with Tucker's (1964, 1966) three-mode factor analysis. Tucker has been the principal advocate of three-dimensional factor analysis, and Tucker (1972) himself relates his work to modern developments of individual scaling models. The last type of analysis relies on comparing grids in pairs and analysing the resulting comparison matrix by some appropriate multidimensional scaling technique (see, for example, Shepard, Romney and Nerlove, 1972). Comparisons between grids may be made in several ways but require detailed information on the performance of the individual grids. The simplest is to replace each grid by its average construct scores and use Pythagoras's theorem to evaluate the distances between two grids. If the relative orientation of the elements in construct-space for each pair of grids is of no interest, then a measure of their agreement can be obtained by translating and then orthogonally rotating one of the pair to make the best fit with the other. This kind of approach is termed 'Procrustes rotation' (see Gower, 1971; Schönemann and Carroll, 1970). Rather then compare pairs of grids, all grids may be rotated simultaneously to fit their joint mean or consensus grid. This I have termed 'Generalised Procrustes analysis' (Gower, 1975).

There is insufficient space to enlarge on these types of analysis here. Instead it will be shown how the analysis of the ordinary two-dimensional grid can be explained in terms of a simple multiplicative model which can be generalized to give an analysis of sets of grids.

12.2 Multiplicative models for single grids

Consider an m by n grid with observations y_{ij} and suppose the multiplicative model (12.1) is to be fitted to these observations:

$$y_{ij} = \mu + \alpha_i + \beta_j + \gamma_i\gamma_j' + \text{error} \quad (i = 1, 2, \ldots, m; \quad j = 1, 2, \ldots, n) \quad (12.1)$$

Denoting means by the dot notation, the identity

$$[\mu + \alpha_i + \beta_j + \gamma_i\gamma_j' \equiv [\mu + \alpha. + \beta. + \gamma.\gamma'.] + [\alpha_i + \gamma'.\gamma_i - \alpha. - \gamma.\gamma'.]$$
$$+ [\beta_j + \gamma.\gamma_j' - \beta. - \gamma.\gamma'.] + [(\gamma_i - \gamma.)(\gamma_j' - \gamma'.)]$$

allows us to assume that (12.1) is in the form where:

$$\Sigma\alpha_i = \Sigma\beta_j = \Sigma\gamma_i = \Sigma\gamma_j' = 0$$

With these convenient, but not essential, restrictions the least squares estimates of α_i, β_j, γ_i and γ_j' are given by:

$$\left.\begin{array}{l} a_i = y_i. - y.. \\ b_j = y._j - y.. \\ \mathbf{Z}c' = C'c \\ \mathbf{Z}^Tc = Cc' \end{array}\right\} \tag{12.2}$$

where $\mathbf{Z}$ is a matrix of residuals with elements

$$z_{ij} = y_{ij} - y_i. - y._j + y..$$

and

$$C = \sum_{i=1}^{m} c_i^2, \, C' = \sum_{j=1}^{n} c_j'^2$$

(This notation is used below, writing a capital letter to denote sums of squares corresponding to the small letter constants, T to denote matrix transposition and the prime symbol for one of a pair of constants, occurring as a product term in a model.) Thus the constants α_i and β_j are estimated in the same way as when fitting main effects to a randomized block experiment. The last two equations of (12.2) may be written in the form:

$$
\begin{aligned}
(\mathbf{Z}^T\mathbf{Z})\,\mathbf{c}' &= CC'\mathbf{c}' \\
(\mathbf{Z}\,\mathbf{Z}^T)\,\mathbf{c} &= CC'\mathbf{c}
\end{aligned}
\right\} \tag{12.3}
$$

showing that the multiplicative terms $\mathbf{c}$ and $\mathbf{c}'$ are the elements of vectors corresponding to the common maximum latent root $\lambda_1 = CC'$ of $\mathbf{Z}^T\mathbf{Z}$ and $\mathbf{Z}\mathbf{Z}^T$. The other latent roots and vectors correspond to additional multiplicative terms in (12.1). The vector scaling is partially restricted, for if $\mathbf{c}$ is arbitrarily scaled so that $C = k$, then $C' = \lambda_1/C = \lambda_1/k$ gives the necessary scaling for C', ensuring that the product $c_i c_j'$ is scale-free. To give equal scaling, choose $k = \lambda_1^{1/2} = C = C'$. The additive components of (12.1) contribute orthogonal elements to analysis of variance of the y_{ij}; the contribution from the pair of multiplicative constants is $CC' = \lambda_1$ and because of the zero sum restrictions and the orthogonality of the latent vectors of a symmetric matrix, successive multiplicative terms will contribute $\lambda_2, \lambda_3, \ldots$, etc. There are $r = \min(m-1, n-1)$ non-zero values of λ.

This analysis of variance gives the table:

Source of variation	Degrees of freedom	Sum of squares
Elements	$m-1$	mA
Constructs	$n-1$	nB
Primary multiplicative terms	t_1	λ_1
Secondary multiplicative terms	t_2	λ_2
Tertiary multiplicative terms	t_3	λ_3
.	.	.
.	.	.
.	.	.
Final multiplicative terms	t_r	λ_r
Total	$nm-1$	$\Sigma(y_{ij} - y_{..})^2$

The exact degrees of freedom to be associated with the multiplicative terms are uncertain but in analogy with approximate chi-squared tests in multivariate canonical analysis, it might be expected that $t_i = n + m - 1 - 2i$ $(i = 1, 2, \ldots, r)$ would be satisfactory. The work of Mandel (1971) who tabulates degrees of freedom obtained from simulation work indicates the degree of approximation involved. Corsten and van Eijnsbergen (1972, 1973) show that under the hypothesis of no multiplicative effect, $\Sigma_{i=1}^k \lambda_i/\sigma^2$ where σ^2, is the *known* variance of the error terms of (12.1), has a standard Wishart distribution with $m-1$ and $n-1$ degrees of freedom. The variance σ^2 may be estimated from $\Sigma_{i=k+1}^r \lambda_i$ but this again introduces approximations.

The model is not new and seems first to have been considered by Fisher and Mackenzie (1923). Gilbert (1963) used it to analyse genetic data, showing that a multiplicative model fitted at least as well as an additive model. This suggested that the classical additivity assumption in genetics was not necessarily so well supported by experimental evidence as previously thought, and similar recent work with agricultural experiments throws some doubt on the universal use of additive interaction terms there. Mandel (1969) has discussed this problem further and relates the multiplicative model to the extensive literature on tests for non-additivity. Gollob (1968) discussed the model in a psychological context, stressing the value of replication. With replication the residual variance σ^2 may be estimated independently of the latent roots leading to more reliable, but still approximate, tests of significance. Gabriel (1971) has found the model useful in the graphical analysis of multivariate data.

What is essentially the same model has been applied to the analysis of two-way contingency tables where it gives rise to *correspondence analysis* (from the French *analyse factorielle des correspondances*). Hill (1974) reviews the several independent discoveries of the method and gives different models that engender the same analysis and computational techniques.

Alternative forms of the last two equations of (12.2) are

$$\begin{pmatrix} C\mathbf{I}_m & \mathbf{Z} \\ \mathbf{Z}^T & C'\mathbf{I}_n \end{pmatrix} \begin{pmatrix} \mathbf{c} \\ \mathbf{c}' \end{pmatrix} = (C + C') \begin{pmatrix} \mathbf{c} \\ \mathbf{c}' \end{pmatrix} \tag{12.4}$$

where $\mathbf{I}_m$ and $\mathbf{I}_n$ are unit matrices of order m and n respectively, and the singular-value decomposition

$$\mathbf{Z}^T = \mathbf{C}'\mathbf{\Sigma}\mathbf{C}^T \tag{12.5}$$

where $\mathbf{C}$ and $\mathbf{C}'$ are orthogonal matrices of all sets of constants $\mathbf{c}$, $\mathbf{c}'$ (now normalized to unit sum of squares) whose correct scaling is given by the diagonal matrix $\mathbf{\Sigma}$ of the square roots of the latent roots of equation (12.3). Singular-value decomposition algorithms are given by Golub (1969) and Golub and Reinsch (1970).

Equation (12.3) establishes a link with principal components analysis, allowing the constructs $\mathbf{c}$ and elements $\mathbf{c}'$ to be plotted in separate spaces. These spaces are linked and it is possible to represent constructs and elements in the same space (Slater, 1972; Hill 1974), which is also one of the prime motives of correspondence analysis and multidimensional unfolding (see, for example, Coombs 1964; Schönemann, 1970). Equation (12.4) can be compared with similar forms occurring in the development of the three-factor multiplicative model extensions (see equation (12.12), below). As mentioned earlier, the singular-value decomposition (12.5) relates to Carroll and Chang's (1970) general canonical decomposition for three (or more)-way tables.

Before leaving two-way grids, it should be clear that there are many variants of the model (12.1). Any of the constant μ, α_i, β_j can be omitted, in which case the simple constraints adopted above are no longer applicable but the least squares equations are easily modified. Another variant, which changes the form of the least squares equations, is when γ_i (or γ'_j) is set to a multiple of α_i (or β_j), but these forms do not seem to give models likely to be of value for analysing repertory grids.

12.3 Three-factor multiplicative models

The model (12.1) can be extended to cope with three-factor versions of Dr. Slater's repertory grid work. The simplest model is to fit y_{ijk} by

$$\hat{y}_{ijk} = m + a_i + b_j + c_k + d_j d'_k + e_i e'_k + f_i f'_j \ (i = 1, 2, \ldots, l;$$

$$j = 1, 2, \ldots, m; \quad k = 1, 2, \ldots, n) \tag{12.6}$$

where the primes have been associated with the highest (lexicographic) ranking subscript in each multiplicative pair. As for the two-factor model, we can always assume that origins have been defined so that

$$\Sigma \, a_i = \Sigma \, b_j = \Sigma \, c_k = \Sigma \, d_j = \Sigma \, d'_k = \Sigma \, e_k = \Sigma \, e'_j = \Sigma \, f_i = \Sigma \, f'_j = 0$$

giving the usual least squares estimates $m = y \ldots$, $a_i = y_i \ldots - y \ldots$, etc., and a residual vector $z_{ijk} = y_{ijk} - m - a_i - b_j - c_k$. The least squares estimates of the multiplicative constants require three matrices, Z_i, Z_j and Z_k. These are the marginal matrices of the residuals obtained by averaging over the named subscript. The normal equations for the multiplicative constants are

$$\left. \begin{array}{ll} Z_i d' = dD', & Z_i^T d = d'D \\ Z_j e' = eE', & Z_j^T e = e'E \\ Z_k f' = fF', & Z_k^T f = f'F \end{array} \right\} \tag{12.7}$$

Thus each pair of sets of multiplicative constants is estimated from the singular-value decomposition of the matrix obtained from the residual vector by averaging over all factors not involved.

If i is the suffix for elements, j for constructs and k for replications (occasions or people), then Z_k is formed by averaging over replicates, and its analysis can be interpreted as that of the average two-way grid. Similar remarks apply to the analysis of Z_i and Z_j, but it is less easy to identify them meaningfully in terms of grids, unless grids of average scores for elements-by-occasions and constructs-by-occasions are useful concepts. This reservation loses its force in other applications where Z_i, Z_j and Z_k all have the same standing.

An orthogonal analysis of variance can be constructed as before, with multiplicative pairs of constants giving additive contributions DD', EE' and FF' plus additional sets, when the vectors corresponding to smaller latent roots are used to estimate additional multiplicative components. Each two-way table has a different set of latent roots. Z_i has $r = \min (m - 1, n - 1)$ roots λ_i ($i = 1, 2, \ldots, r$), Z_j has $s = \min (l - 1, n - 1)$ roots μ_j ($j = 1, 2, \ldots, s$) and Z_k has $t = \min (l - 1, m - 1)$ roots ν_k ($k = 1, 2, \ldots, t$). The analysis of variance therefore takes the following form:

Source of variation	Degrees of freedom	Sum of squares
Elements (E)	$l-1$	$mn\sum (y_{i..}-y...)^2$
Constructs (C)	$m-1$	$ln\sum (y._{j.}-y...)^2$
Occasions (O)	$n-1$	$lm\sum (y.._k-y...)^2$
C x O interaction	$(m-1)(n-1)$	$\begin{aligned}&\lambda_1 l\\&\lambda_2 l\\&\vdots\\&\lambda_r l\end{aligned}$
E x O interaction	$(l-1)(n-1)$	$\begin{aligned}&\mu_1 m\\&\mu_2 m\\&\vdots\\&\mu_s m\end{aligned}$
E x C interaction	$(l-1)(m-1)$	$\begin{aligned}&\nu_1 n\\&\nu_2 n\\&\vdots\\&\nu_t n\end{aligned}$
Total	$lmn-1$	$(y_{ijk}-y...)^2$

No formal work seems to have been done on the degrees of freedom to be attached to the individual terms corresponding to single sets of multiplicative constants in this analysis of variance, but as each interaction sum of squares is derived from the analysis of a two-way average grid the breakdown of degrees of freedom previously discussed for the two-way table should be applicable. Mandel's (1971) work on degrees of freedom in the two-way case should be appropriate here also. Thus the terms $\lambda_1, \lambda_2, \ldots, \lambda_r$ are derived from the singular-value decomposition of the two-way table of residuals Z_i, or alternatively as the latent roots of $Z_i Z_i^T$, and similarly for the μ's and ν's. Thus the two-factor model has a simple generalization to three factors, and clearly four or more factors could be dealt with in the same way.

However, another form of generalization to three factors is to include three-factor products like $u_i v_j w_k$, and it is shown below how this leads to certain difficulties. The identity

$$uvw \equiv (u-u.)(v-v.)(w-w.) + u.(v-v.)(w-w.) + v.(u-u.)(w-w.)$$

$$+ w.(u-u.)(v-v.) - v.w.(u-u.) - u.w.(v-v.)$$

$$- u.v.(w-w.) + u.v.w. \tag{12.8}$$

includes terms like $v.(u-u.)(w-w.)$ on the right-hand side, and this shows that if we assume $\sum u_i = \sum v_j = \sum w_k = 0$ we must for consistency also include terms proportional to $v_j w_k$ in the model. Defining

$$
\left. \begin{array}{l}
u_i' = \dfrac{u_i - u_.}{u_.}\ (u_.\ v_.\ w_.)^{1/2} \\[3mm]
v_j' = \dfrac{v_j - v_.}{v_.}\ (u_.\ v_.\ w_.)^{1/2} \\[3mm]
w_k' = \dfrac{w_k - w_.}{w_.}\ (u_.\ v_.\ w_.)^{1/2}
\end{array} \right\}
$$

shows that (dropping the primes) the simplest model of this kind that can be considered for three factor products is

$$
\hat{y}_{ijk} = m + a_i + b_j + c_k + v_j w_k + u_i w_k + u_i v_j + \rho u_i v_j w_k \tag{12.9}
$$

where $\Sigma a_i = \Sigma b_j = \Sigma c_k = \Sigma u_i = \Sigma v_j = \Sigma w_k = 0$ and ρ is a constant to be estimated. This model gives an additive analysis of variance with component sums of squares $A, B, C, VW, UW, UV, \rho^2 UVW$. Although it appears to include two-factor product terms, these are merely a consequence of the zero-sum restrictions and are not independent of the triple-product term. The mean and main effects have the same estimates as before, giving the same residuals z_{ijk}. Recalling the previous notation, $\Sigma u_i^2 = U$, etc., minimizing with respect to ρ gives

$$
\rho UVW = \sum_{i,j,k} z_{ijk} u_i v_j w_k \tag{12.10}
$$

and minimizing with respect to u, v and w gives

$$
\left. \begin{array}{l}
Z_j w + Z_k v + p = u(\rho^2 VW + V + W) \\[2mm]
Z_i w + Z_k^T u + q = v(\rho^2 UW + U + W) \\[2mm]
Z_i^T v + Z_j^T u + r = w(\rho^2 UV + U + V)
\end{array} \right\} \tag{12.11}
$$

where p is the vector with elements

$$
\rho(v^T Z_i w - VW) \quad (i = 1, 2, \ldots, l)
$$

q is the vector with elements

$$
\rho(u^T Z_k w - UW) \quad (j = 1, 2, \ldots, m)
$$

and r is the vector with elements

$$
\rho(u^T Z_k v - UV) \quad (k = 1, 2, \ldots, n)
$$

In general these equations are not easy to solve, but when $\rho = 0$ we have a special case of the two-factor product model with which some progress can be made and which has interesting links with the general two-dimensional case. With $\rho = 0$ (12.11) becomes

$$
\begin{bmatrix} UI_l & Z_k & Z_j \\ Z_k^T & VI_m & Z_i \\ Z_j^T & Z_i^T & WI_n \end{bmatrix} \begin{bmatrix} u \\ v \\ w \end{bmatrix} = (U + V + W) \begin{bmatrix} u \\ v \\ w \end{bmatrix} \tag{12.12}
$$

where I_l, etc., are unit matrices of order l, etc. Thus $(\mathbf{u}^T, \mathbf{v}^T, \mathbf{w}^T)^T$ is the latent vector of the matrix on the left-hand side of equation (12.12) corresponding to the largest latent root $\lambda_1 = (U + V + W)$ and scaled so that the sum of squares of its elements is λ_1. This is a variant of the usual latent-root-and-vector problem, because U, V and W occur on the left-hand side of equation (12.12) and are reminiscent of the role of communalities in factor analysis. This shows that only the vector with maximum $(U + V + W)$ is of interest.

The solution of (12.12) may be regarded as a three-dimensional form of singular-value decomposition corresponding to the two-dimensional form given in equation (12.4). A possible method of solution is to take initial estimates U_0, V_0, W_0 of U, V and W, and insert these in the diagonals of the matrix on the left-hand side of equation (12.12), finding the principal root λ_0 and corresponding vector $(\mathbf{u}_0, \mathbf{v}_0, \mathbf{w}_0)$ scaled to have sum of squares λ_0. The sum of squares of the first l terms of the vector give U_1, the next m terms give V_1 and the final n give W_1, the approximations to start a second round of iteration. Iteration proceeds until U, V, W and $(\mathbf{u}, \mathbf{v}, \mathbf{w})$ converge. The initial estimates may be taken from the two-way analyses of Z_i, Z_j and Z_k; their first latent roots λ_i, μ_j and ν_k are (respectively) estimates of VW, UW and UV. Hence estimate U_0 by $(\mu_1 \nu_1 / \lambda_1)^{1/2}$, etc.

When $\rho \neq 0$ in (12.11) the solution is more difficult. One possibility, which can also be used when $\rho = 0$, is to get initial estimates as previously and insert these in equation (12.10) to give ρ_0. These values can now be inserted in (12.11) to give the next estimates, $\mathbf{u}'$ $\mathbf{v}'$ and $\mathbf{w}'$, as follows:

$$\mathbf{u}' = \frac{Z_j \mathbf{w} + Z_k \mathbf{v} + \mathbf{p}}{\rho^2 \, VW + V + W}, \quad \text{etc.}$$

The methods of solution proposed may be compared with iterations on an initial estimate of a latent vector (Aitken, 1937) where the rate of convergence depends on the ratio of the first two latent roots. I tried numerical procedures of this kind with varying success. Sometimes convergence is satisfactory, at others not. Recalling the amount of work that has been put into getting satisfactory procedures for the ordinary algebraic eigenvalue problem (Wilkinson, 1965) and how long it has taken to get satisfactory numerical methods for solving the maximum likelihood factor analysis estimation equations (Joreskog, 1967), it is hardly surprising that difficulties arise with the above methods.

If vectors with smaller roots are included as additional three-factor product terms in the model (9), the additive analysis of variance property is not preserved. This remark is equivalent to that of Carroll and Chang (1970, p. 312) who write:

> One thing that has become evident to the present authors is that the three-way or higher-way case is inherently more complicated than the two-way case in at least one important aspect. This is that, while in the decomposition of two-way tables it is possible to proceed 'one dimension at a time', this is not, in general, possible in higher-way cases. By this we mean that in the two-way case it is always possible to do a one-dimensional analysis (that is, for $r = 1$), calculate a 'residual' matrix, do a second one-dimensional analysis of the 'residual' matrix, and so on r times, combining the results of the r separate one-dimensional analyses to form a single r-dimensional analysis. This is not true in the three-way or higher-way case.

Carroll and Chang's model differs from those discussed above, and, in the three-way case, is

$$\hat{y}_{ijk} = u_i v_j w_k \qquad (12.13)$$

This is simpler than the models discussed previously because it omits the additive terms a_i, b_j and c_k. However, if zero-sum constraints are put on u_i, v_j and w_k of equation (12.13), then this induces extra terms, as in the expansion (12.8), which are not helpful in getting a simple least squares solution. Hence Carroll and Chang fit (12.13) directly, choosing initial values of v_j and w_k and estimating u_i by a least squares (regression) fit. Then v_j is fitted using the new values of u_i and the current values of w_k. The iterative procedure continues estimating new values of w_k and then repeating the cycle until the values have converged. Carroll and Chang report that this process converges satisfactorily, although they cannot show that it must always do so or that solutions arrived at in this way are unique. Clearly this kind of calculation has much in common with the methods outlined above for solving the equations (12.11) and (12.12); in fact the vectors p, q and r are multiples of the regression coefficients in Carroll and Chang's approach. In the two-dimensional case Carroll and Chang's technique is exactly that of correspondence analysis (see Hill, 1974). There the mathematical problem is that of singular-value decomposition which is closely related to standard matrix eigenvalue problems. There is an extensive numerical analytical literature for this (see Wilkinson, 1965), and the proper numerical investigation of the problems discussed here needs similar investigation.

The difficulties met with the three-factor models so far discussed become even more pressing with the most general three-factor model, which is a combination of (12.5) and (12.9):

$$\hat{y}_{ijk} = m + a_i + b_j + c_k + d_j d_k' + e_i e_k' + f_i f_j' + v_j w_k + u_i w_k + u_i v_j + \rho u_i v_j w_k$$

where $\hspace{8cm} (12.14)$

$$\Sigma a_i = \Sigma b_j = \Sigma c_k = \Sigma d_j = \Sigma d_k' = \Sigma e_i = \Sigma e_k' = \Sigma f_i = \Sigma f_j' = \Sigma u_i$$
$$= \Sigma v_j = \Sigma w_k = 0$$

The two-product terms occur in pairs like $(d_j d_k' + v_j w_k)$ because the above restrictions impose the component $v_j w_k$ through identity (12.8) as well as the 'genuine' component $d_j d_k'$. This implies that the analysis of variance is no longer additive in orthogonal components, for product terms like $\Sigma_{i,j,k} [(d_j v_j)(d_k' w_k)]$ do not vanish. The main effects and residuals are estimated as before and (12.10) is still the equation for ρ. For d, e, f, d', e', f' we now have

$$Z_i d' = dD' + v \Sigma (w_k d_k'), \qquad Z_i^T d = d'D + w \Sigma (v_j d_j), \text{ etc.} \qquad (12.15)$$

and for u, v, w,

$$Z_j w + Z_k v + p' = u (\rho^2 VW + V + W) + e \Sigma (w_k e_k') + f \Sigma (v_j f_j'), \text{ etc.} \qquad (12.16)$$

where p' is the vector with elements

$$\rho [v^T Z_i w - VW - \Sigma (v_j d_j) \Sigma (w_k d_k')], \text{ etc.}$$

I have not attempted to use equations (12.15) and (12.16) or think of ways they might be solved, but it is clear that such three-factor-product models require much numerical analytical work to be done before they can be fitted confidently.

12.4 Conclusion

We have seen how the analysis of a simple two-way grid can be expressed as fitting, by least squares, a multiplicative model. The extension to three-way models gives rise to both methodological and computational problems. Many different extensions are possible. The model (12.6) involving three-factors but with only 'two-factor interactions', leads to tractable computations of the same kind as are required for two-way grids. However, the special case (12.9) of the 'three-factor interaction' model (12.12) already introduces computational problems which are even more severe with (12.12) itself. The full three-factor model (12.14) is more complex still and is little investigated. Carroll and Chang's (1970) general canonical decomposition, in its three-factor forms, gives a simple model with a practicable computational process. However, it does not include additive terms (written as a_i, b_j and c_k in the above), which would seem to be essential when the observations are measured on certain types of scale. A different aspect of grid comparisons is given by analysing the set of residual sum-of-squares values obtained from the Procrustes rotation of all pairs of grids, or by rotating all grids simultaneously.

What is required now is an attack on the two fronts. Psychologists must try out and evaluate the models discussed here (and those briefly mentioned in the introduction) to decide which are appropriate for analysing three-way grids under different conditions. Already some of these models can be fitted without difficulty, but others require extensive investigation by numerical analysts before satisfactory numerical algorithms can be produced.

Acknowledgement

Some of this work was done while the author was visiting the Department of Biostatistics, University of North Carolina, at Chapel Hill in 1970, where it was supported by the National Institutes of Health, Institute of General Medical Sciences, Grant No. 12868—07.

References

Aitken, A. C. (1937). 'The evaluation of the latent roots and vectors of a matrix'. *Proc. roy. Soc. Edinb.*, 57, 269—304.
Carroll, J. D., and Chang, J. J. (1970). 'Analysis of individual differences in multidimensional scaling via an *n*-way generalization of 'Eckart—Young' decomposition'. *Psychometrika*, 35, 283—319.
Coombs, C. H. (1964). *A Theory of Data*, John Wiley and Sons, New York.
Corsten, L. C. A., and van Eijnsbergen, A. C. (1972). 'Multiplicative effects in two-way analysis of variance'. *Statistica Neerlandica*, 26, 61—68.
Corsten L. C. A., and van Eijnsbergen, A. C. (1973). Addendum to 'Multiplicative effects in two-way analysis of variance'. *Statistica Neerlandica*, 27, 51.
Eckart, C., and Young, G. (1936). 'The approximation of one matrix by another of lower rank'. *Psychometrika*, 1, 211—218.

Fisher, R. A., and Mackenzie, W. A. (1923). 'Studies in crop variation. II. The manurial response of different potato varieties'. *J. agric. Sci.*, 13, 311—320.

Gabriel, K. R. (1971). 'The biplot graphic display of matrices with applications to principal components'. *Biometrika*, 58, 453—467.

Gilbert, N. (1963). 'Non additive combining abilities'. *Genet. Res.*, 4, 65—73.

Gollob, H. F. (1968). 'A statistical model which combines features of factor analytic and analysis of variance techniques'. *Psychometrika*, 33, 73—115.

Golub, G. H. (1969). 'Matrix decomposition and statistical calculations'. In R. C. Milton and J. A. Nelder (Eds.), *Statistical Computation*, Academic Press, New York and London, pp. 365—397.

Golub, G. H., and Reinsch, C. (1970). 'Handbook series linear algebra, singular value decomposition and least squares solutions'. *Numer. Math.*, 14, 403—420.

Gower, J. C. (1971). 'Statistical methods of comparing different multivariate analyses of the same data'. In F. R. Hodson, D. G. Kendall and P. Tartu (Eds.), *Mathematics in the Archaeological and Historical Sciences*, The University Press, Edinburgh.

Gower, J. C. (1975). 'Generalised Procrustes Analysis'. *Psychometrika*, 40, 33—51.

Hill, M. O. (1974). 'Correspondence analysis: a neglected multivariate method'. *J. R. Statist. Soc.(C)*, 23, 340—354.

Joreskog, K. G. (1967). 'Some contributions to maximum likelihood factor analysis'. *Psychometrika*, 32, 443—482.

Mandel, J. (1969). 'The partitioning of interaction in analysis of variance'. *J. Res. Nat. Bur. Stand. (U.S.)*, 73B, 309—328.

Mandel, J. (1971). 'A new analysis of variance model for non-additive data'. *Technometrics*, 13, 1—18.

Schönemann, P. H. (1970). 'On metric multidimensional unfolding'. *Psychometrika*, 35, 349—366.

Schönemann, P. H., and Carroll, R. M. (1970). 'Fitting one matrix to another under choice of central dilation and a rigid motion'. *Psychometrika*, 35 , 245—255.

Shepard, R. N., Romney, A. K. and Nerlove, S. B. (1972). *Multidimensional Scaling*, Vol. I. Theory, Vol. II. Applications. Seminar Press, New York.

Slater, P. (1972). Composite diagrams an systems of angular relationships applying to grids. *Paper presented to the Royal Statistical Society, multivarate analysis study group.*

Tucker, L. R. (1964). 'The extension of factor analysis to three-dimensional matrices'. In N. Frederiksen and H. Gulliksen, (Eds.) *Mathematical Psychology*, Holt. Rinehart and Winston, New York. pp. 109—127.

Tucker, L. R. (1966). 'Some mathematical notes on three-mode factor analysis. *Psychometrika*'. 31, 279—311.

Tucker, L. R. (1972). 'Relations between multidimensional scaling and three-mode factor analysis'. *Psychometrika*, 37, 3—27.

Wilkinson, H. H. (1965). *The Algebraic Eigenvalue Problem*, Oxford University Press, London.

13

THE PSYCHOLOGICAL MEANING OF
STRUCTURAL MEASURES DERIVED
FROM GRIDS

Jane Chetwynd

St. George's Hospital Medical School

In Chapter 11 the concept of structural features present in grids has been introduced and the limited use of such variables as a means of comparing one grid with another discussed.

This chapter describes attempts to investigate the psychological significance, if any, of three structural grid measures which are obtained with the INGRID program:

(a) measure of the cognitive complexity/simplicity of the informant given by the explanation power of the first component of the grid.
(b) Measures of response error:
   (1) Bias — a measure of the tendency of an informant to respond by using only one end of the grading scale.
   (2) Variability — a measure of the tendency of an informant to respond by using the extremes of the grading scale.

If such variables can be seen to have some psychological meaning then their use extends beyond simply a comparison medium to that of a measure of the psychological significance of the grid and so to an insight into personality and/or cognitive features of the informant.

Firstly, in this chapter the three measures will be described in terms of the phenomena which they attempt to assess, and relevant findings concerning the correlates of such phenomena will be given. Secondly, a report will be presented of three investigations which attempted to examine various personality concomitants of the three structural measures.

13.1 The structural measures and their correlates

(a) A measure of cognitive complexity/simplicity given by explanation power of the first component

It was suggested by Bannister and Mair (1968) that the most extensive use of repertory grids outside the theoretical framework in which Kelly surrounded the technique had been in the measurement of the postulated dimension of cognitive complexity. This statement is hardly applicable now as grids are being used in many and varied situations (see Chetwynd, 1975), but it remains an important feature of grids that they can be viewed in this way.

The grid was first put to this use by Jones (1954) who assessed cognitive complexity by the 'explanation power' of the first factor extracted from the rep grid and also by the number of factors that could be derived from the rep grid.

In the INGRID analysis the explanation power of the first component is given by the percentage of the total variation accounted for by the first component, and so this can be considered an inverse measure of cognitive complexity. The recent rapid growth of interest and research into cognitive complexity as a personality variable is due to the enthusiasm of James Bieri, a student of Kelly, who developed the concept within the general context of Kelly's personal construct theory (Bieri, 1955, 1961). Bieri (1955) defines cognitive complexity in terms of the versatility of the individual's construct system: 'A system of constructs which differentiate highly among persons is considered to be cognitively complex. A construct system which provides poor differentiation among persons is considered to be cognitively simple in structure.'

Since this description by Bieri, a number of investigators have refined or broadened the definition of cognitive complexity so that we now have many different phenomena labelled cognitive complexity/simplicity and many differing ways of measuring those phenomena (e.g. Crockett, 1965; Stringer, 1971; Vannoy, 1965). It is perhaps typical of psychologists to be attracted to a new concept and develop and extend it so that it becomes such an overinclusive term that it encompasses many more concepts than was its original designation.

The varieties of definitions of cognitive complexity are reflected in the variety of measures which have been used to evaluate it (e.g. Bieri and Blacker, 1956; Binner, 1958; Pederson, 1958; Tripodi and Bieri, 1963, 1964; Warren, 1964). A number of studies have been carried out to investigate the relationship, if any, between these various measures, but the results generally suggest little in common between them (e.g. Gault, 1971; Hall, 1966; Pederson, 1958; Sechrest and Jackson, 1961; Vannoy, 1965).

Correlates of cognitive complexity   Collating the research concerning the correlates of cognitive complexity is difficult because of the variety of measures of complexity which have been used. A brief survey will be made here of the relevant research findings of importance, but more detailed reports can be found in Bieri (1961) and Crockett (1965).

The focus of research on cognitive complexity has been in the field of interpersonal or social behaviour. Bieri (1955) initiated this interest by studying the relationship between cognitive complexity and the prediction of the behaviour of others. The findings were that the more cognitively complex a person the more able he was in his predictions of others. However, in subsequent studies, e.g. Leventhal (1957) and Sechrest and Jackson (1961), although a positive relationship between complexity and accuracy in prediction was also found, the results did not reach significance.

A further finding of Bieri's (1955) study was that subjects low in complexity showed greater assimilative projection, i.e. greater expectation of similarity in attitudes of self and others. Campbell (1960) also reported that subjects low in cognitive complexity assume others who they like to be more similar to themselves than do those high in complexity. Further findings of Campbell's study concerned the awareness of positive and negative attributes in others. Cognitively simple subjects showed a greater tendency to divide people in a good-bad dichotomy, and they assumed greater mutuality of liking and disliking within their social group. The results of both the studies of Scott (1963) and that of Supnick (1964) provide evidence that people high in cognitive complexity are more likely to use both positive and negative terms in their descriptions of others.

Several studies have been concerned with the relationship between cognitive complexity and the resolution of contradictory information in forming an impression. Crockett (1965), Lo Giudice (1963), Mayo (1959), Mayo and Crockett (1964) and Nidorf (1961) showed that cognitively complex persons are better able to form an integrated impression of others when presented with both negative and positive information about them. As a consequence of this, Mayo and Crockett (1964) showed that cognitively complex persons are less affected by the order of presentation in forming impressions of other people.

In a study of attitude change, Lundy and Berkowitz (1957) found that the most cognitively simple subjects exhibited the least change after a change-inducing experience, while the most complex subjects exhibited negative change, i.e. they increased the level of their original attitudes. Those who were neither extremely simple nor complex were consistently susceptible to change. In the field of probability preferences, Higgins (1959) showed that cognitively complex people tend to prefer moderate probabilities in indeterminate events, complexity being assessed in terms of the number of constructs generated.

An interesting finding concerning cognitive complexity and levels of violence was that of Topçu (1976). Working with aggressive patients he found that the more aggressive subjects exhibited more cognitively simple systems, whereas the less aggressive control group were more cognitively complex. Suicide attempters were also found to be more cognitively simple than the controls.

Work on the relationship between cognitive complexity and other personality variables outside the range of interpersonal behaviour has been less extensive.

It seems reasonable to suppose that developmental and experiential differences may be important to the level of complexity. Hunt (1962) in a study of adolescent boys showed that cognitive complexity increases with age and this finding was confirmed with children aged nine to sixteen by Signell (1966). An interesting result of Runkel and Damrin (1961) showed a U-shaped relationship between training and cognitive complexity as measured by the number of constructs used. Trainee teachers were asked about problems of teaching at three stages during their training; on entering the course halfway through and at the end. The U-shape curve concerning their cognitive complexity can be interpreted as follows. At the beginning of the course the trainee has a relatively large number of constructs in his system, but these become adapted as the course progresses and he concentrates on relatively few important areas. By the end of the course he has developed a large number of constructs again relevant to his new area of interest. In contrast, Koch (1958) found no change in cognitive complexity after a concept attainment experience.

Intelligence, too, could be assumed to have a direct relationship with cognitive complexity. Such a result was found by Hunt (1962) and also by Vannoy (1965).

However, no significant relationship between cognitive complexity and intelligence was found in the studies of Lo Giudice (1963), Mayo (1959), Rosenkrantz (1961) and Sechrest and Jackson (1961).

Few studies have reported sex differences in the level of cognitive complexity. Hall (1966), in a study of 110 subjects completing five different grids yielding five different complexity scores, reported a significant difference between the total complexity scores for men and women, the men being the more complex. There were also differences in the correlations between the five complexity scores for the men and women. Eight of the ten intercorrelations were significant for men, while none were for the women. In contrast, Higgins (1959) reports that females were significantly more complex than males in his study of probability preferences.

Finally, the results concerning the relationships between cognitive complexity and the three important personality variables — psychoticism, extraversion and neuroticism — must be mentioned. Jones (1954) found that neuropsychiatric patients had factorially more simple cognitive structures than normal control subjects. Bannister (1960, 1962, 1963, 1965) has shown repeatedly that schizophrenic patients have lower intensity of relationship scores than normal persons or neurotic or depressive patients, though the interpretation of these findings has been much criticized (Frith and Lillie, 1972; Haynes and Phillips, 1972; Jaspars, 1966). In terms of the introversion/extraversion continuum, Bieri and Messerley (1957) found that extraverts were verbally more complex than introverts.

The many contradictory and confusing findings in this area must be blamed in part on the lack of congruity of the measures of complexity used. However, it is argued by some workers, e.g. Bannister and Mair (1968), that the concept of cognitive complexity/simplicity as a personality trait and a stable fixed quantity is incorrect, and they cite such discrepant findings as proof of their arguments.

(b) Measures of response error — bias and variability

Grid technique, in common with any form of questionnaire method which attempts to assess attitudes and opinions, may be subject to various forms of response error. Such errors have variously been described as response style, response set, response bias, etc. They are simply consistent patterns of response behaviour, exhibited in the questionnaire completion situation, which reflect personality characteristics of the respondent rather than responses to the questions. As such, it is thought that they contaminate the 'true' response and, if possible, their effects should be isolated and eliminated from the data. Apart from their obvious deleterious effects, however, they are interesting to examine as phenomena in their own right, and a further advantage of their isolation is their resultant accessibility for investigation.

The three main forms of response error which attack the data obtained with grid technique are social desirability, extreme response style and a form of acquiescence. Of these the most difficult to assess is social desirability. The extent to which it affects the data is dependent on the elements and constructs which are included in the grid and on the context in which it is administered. The experimenter may be able to control to some degree the effect of social desirability on his subject by careful instruction and construction of the test situation. Furthermore, the results of the grid analysis may help him identify to what extent his subject is likely to have been affected by the need to make socially desirable responses, again dependent on whether that sort of information has been included in the constructs or elements of the grid.

Only those examples of grid technique which employ the grading method of scoring are affected by the acquiescent and extreme response styles.

The concept of acquiescent response style originated in true/false option questionnaires when it was observed that some subjects were more likely to give a 'true' response when they were uncertain of the correct response (acquiescence) and some were more likely to give a 'false' response (criticalness). This concept is extended in Kelly's (1955) dichotomous repertory grid test to the preference of a subject for attributing elements to one rather than the other pole of each of the constructs, and it is further extended in the situation where a larger grading scale is used to the tendency of the subject to place the elements at, or near, one of the poles of each of the constructs.

The extreme response style (ERS) is exhibited by a subject who, when using a grading scale, displays a tendency to attribute to the elements grades which represent either of the polar extremities of the constructs. In other words, rather than assign grades representing the intermediate points on the scale, he tends to use grades to a disproportionate extent, which indicate either of the extremes.

These two styles are therefore related in terms of the amount of discrimination between the elements which is produced. In one there is a great deal of discrimination because elements are attributed grades at both extremes of the scale (ERS); in the other the elements tend to be placed on the same polar side of the constructs and hence little discrimination between them occurs.

This section is concerned with these latter two forms of response error. Measures of both were incorporated into the INGRID program and they are described in Chapter 7: bias is a measure of acquiescence and variablility a measure of extreme responding. As has been noted in that chapter, the links between acquiescence as originally defined and the measure of bias included in the INGRID program are perhaps tenuous, and the generalization of the research findings from one to the other should be made with caution.

Measures of bias and variability have been extracted for quasis, and the evidence available (Chetwynd and Slater, 1976) suggests that bias is very much higher with experimental grids than with quasis and variability is slightly, but significantly, lower. It seems reasonable to suppose from this that there are some psychological implications in these structural measures.

Correlates of acquiescence   Findings concerning the personality correlates of acquiescence are not on the whole conclusive. Perhaps the most consistently reported relationship is the inverse one between acquiescence and measures of intellectual ability (e.g. Elliott, 1961; Forehand, 1962; Shaw, 1961). Negative relationships between acquiescence and verbal knowledge, general reasoning and deductive thinking have been shown by Messick and Frederiksen (1958). Jackson and Pacine (1961) showed that acquiescence was negatively related to college results, to verbal knowledge and to general reasoning ability. Cattell, Dubrin and Saunders (1954) found relationships between acquiescence and low verbal ability and poor performance in riddle solving. The general conclusions from many reports reflect a stable relationship between acquiescence and certain cognitive variables.

The study of Couch and Keniston (1960) showed relationships between acquiescence and various personality traits. High acquiescence was associated with measures of impulsivity. dependency, anxiety, mania anal resentment and anal preoccupation; low acquiescence was associated with ego-strength, stability, responsibility, tolerance and impulse control. Similar findings are those of Cattell (1955) and Cattell and Gruen (1955), who showed associations between measures of acquiescence and high speed of judgement, reaction in perceptual and perceptual-motor tasks, rapid tempo in preferred rates of movement and, to some extent, with high verbal fluency.

Couch and Keniston (1960) were also able to obtain certain clinical variable correlates of acquiescence. They found high acquiescence exhibitors were typically impulsive, emotionally reactive, extraverted, externally orientated, low in psychological inertia and possessed passive egos. On the other hand, low acquiescence exhibitors were guarded, defensive, constricted, inhibited, introverted, withdrawing, introspective, high in psychological inertia, slow and critical reactors, and possessed active egos.

Messick and Kogan (1967) found relationships between acquiescence and impulsiveness, the admission of undesirable characteristics and anxiety symptoms. They were also able to infer tentatively that acquiescence was more pervasive in women than in men — 'not because mean levels of acquiescence are higher in females, which is generally not the case, but because more variables are directly implicated for females as correlates of acquiescence'.

The overall picture of the personality correlates of the acquiescent response style obtained from the reported research can be collapsed into two basic factors. The first is concerned with the negative relationship between acquiescence and cognitive abilities, of which there is a reasonable and diverse amount of evidence. The second, which is supported in the main by the work of Couch and Keniston (1960), is the relationship between acquiescence and a dimension of impulsiveness. The validity of this latter relationship is perhaps uncertain. Couch and Keniston produced their results with a substantial amount of inference and abstractions, and, as McGee (1962) points out, the deductive process is invaluable in hypothesis building, but the research remains unfinished unless the inductive testing of such hypotheses is completed. This does not appear to have been done to any satisfactory degree.

On the whole there is not a substantial amount of evidence concerning the concomitants of acquiescence and perhaps this fact in itself makes the existence of the concept as a personality variable less credible.

Correlates of extreme response style (ERS)   An excellent review concerning personality attributes associated with extreme response style is given by Hamilton (1968), and this should be referred to for a greater detailed and wider coverage of the subject.

The findings concerning the relationship between age and ERS indicate a curvilinear relationship showing high levels of ERS both in early childhood and in later life. Hesterley (1963) found that adults in an age group ranging between 60 and 83 tended to make more extreme responses that those in the 20–59 age group, as did children in the 6–17 age group. However, there was no significant difference between the older group of adults and the children, and it was observed that the ERS scores of the elderly group approximated those of the children aged 9–10. A number of other studies have shown that ERS decreases throughout childhood and adolescence from a high level in the early years (Light, Zax and Gardiner, 1965; Soueif, 1958; Zax, Gardiner and Lowy, 1964).

These findings indicate the possibility of a relationship between intellectual functioning and ERS. Kerrick (1956) reports finding an inverse relationship between I.Q. and the tendency to give extreme responses in college students, and this was confirmed with the children studied by Light, Zax and Gardiner (1965) and by Stricker and Zax (1966). Further evidence is provided by Wilkinson (1970), who found a similar relationship in alcoholic patients. On the other hand, Zuckerman and Norton (1961), using student nurses as subjects, reported a near-zero correlation between ERS and intelligence.

Differences between males and females in the tendency to give extreme responses have been reported in some studies, though again conflicting findings occur. Many studies have found females to make significantly more extreme responses than males (e.g. Borgatta and Glass, 1961; Brown, 1964; Soueif, 1958). In one study by Brengelmann (1959) males made significantly more extreme responses than females. However, many studies failed to find a significant difference at all between the sexes (e.g. Berg and Collier 1953; Brengelmann, 1960b; Light, Zax and Gardiner, 1965).

The relationships between pathological features of personality and ERS are perhaps the most important in this area. O'Donovan (1965) in a review of the many pathological correlates of ERS mentions the following: neuroses, primitive id impulses, mental-patient status, deviance, maladjustment, anxiety, intolerance of ambiguity, inflexibility, desire for certainty, ethnocentrism, rigidity and dogmatism. It is generally hypothesized that psychiatric patients tend to give more extreme responses than normal subjects and there is a large body of research providing evidence of this (e.g. Barnes, 1955; Brengelmann, 1958, 1959, 1960a, 1960b; Love and Priest, 1967; Neuringer, 1961, 1963; Parsonson, 1969; Zax, Gardiner and Lowy, 1964; etc.). However, some studies have failed to support the hypothesis (e.g. Blumenthal, 1968; Bopp, 1955; Luris, 1959; Zax, Loiselle and Karras, 1960).

There are some reports of differences between the various diagnostic categories of illness in the tendency to make extreme responses. For example, Arthur (1966) found psychotic patients exhibited greater ERS than neurotic patients, Neuringer (1961) found that suicidal neurotics and psychosomatic patients exceeded general medical and surgical patients, and further support is provided by the work of Dresser (1969), Jones (1956) and Wertheimer and McKinney (1952). The general suggestion from these findings is that ERS tends to increase as the degree of pathology increases.

An interesting study by Iwawaki and Zax (1969) linked ERS with personality dimensions and pathology as measured by the Maudsley Personality Inventory (M.P.I.) on a sample of male college students. They found that subjects scoring high on the neurotic dimension made significantly more extreme responses than low scorers and that low neuroticism-scoring introverts had significantly fewer extreme ratings than high neuroticism-scoring extraverts. The difference between the introverts and extraverts as a whole was not quite significant, but was in the direction of extraverts tending to give more ERS responses than introverts.

There have been a number of studies relating ERS to scores on the Taylor Manifest Anxiety Scale. High anxiety subjects made significantly more extreme responses in the studies of Berg and Collier (1953) and of Lewis and Taylor (1955). This finding was confirmed by Norman (1969) who showed 'anxious' subjects made more extreme responses than 'adjusted' subjects when the Welsh A scale of the Minnesota Multiphasic Personality Inventory (M.M.P.I.) was used to measure anxiety.

Finally, the relationship between ERS and authoritarianism should be mentioned. Mogar (1960) showed that high scorers on the F scale tended to use the extremes more often than middle or low F scorers. A finding which is perhaps related is that of Osgood, Suci and Tannenbaum (1957) concerning the tendency of American Legionnaires to use the extreme scores on the Semantic Differential. In contrast, Zuckerman, Norton and Sprague (1958) found significant negative correlation between ERS and authoritarianism and non-significant relationships are reported by Peak, Muney and Clay (1960) and by White and Harvey (1965).

Overall, there appears to be little agreement between the research findings concerning the concomitants of ERS. Perhaps the only general conclusions that can be reached are that ERS increases with increasing degree of pathology and decreases with increasing cognitive functioning and maturity.

3.2 Investigations into the personality concomitants of three structural features of grids — cognitive complexity, bias and variability

In view of the somewhat contradictory findings concerning the personality correlates of cognitive complexity, acquiescence and extreme response style, it was felt necessary to cover a wide variety of measures of personality attributes in this investigation. Hence the study reported here combines observations from three separate experiments, each employing grids from which the various structural measures could be extracted, but each including differing measures of personality variables. In the following methods and results sections each study will be considered separately, but in the discussion section the results from all three studies will be combined and discussed in terms of the overall picture obtained,

Method

Part 1 Forty male students, age range 18—25, were the subjects in this experiment. All subjects were administered the Personality Inventory (P.I.) (Eysenck and Eysenck, 1968a, 1968b, 1971) which is an adaption of the Eysenck Personality Inventory (E.P.I.) (Eysenck and Eysenck, 1964), devised to provide a measure of psychotism as well as measures of extraversion and neuroticism The extraversion score is made up from the total of two separate measures: impulsivity and sociability. Also administered was a second questionnaire within which was embedded a twenty-one item lie scale (Eysenck and Eysenck, 1964). Only the lie scale from this second questionnaire was used for this study.

Each subject completed a grid by grading six females physiques on ten constructs using a seven-point scale. The grid is a standard one used in a number of studies of the stereotyping process and described in detail elsewhere (Chetwynd, 1976).

The grids were analysed by the INGRID program and the following three measures extracted:

(a) The percentage of variation accounted for by the first component in each case to give an inverse measure of cognitive complexity for each subject.

(b) The bias score to give an indication of the subject's tendency to display acquiescence in responding.

(c) The variability score to give a measure of extreme responding.

Product-moment correlations were then calculated between each of these three structural measures and the other variables included in the study.

Part 2 Ninety-five male prisoners, age range 20—48, were the subjects in this experiment. The following questionnaires were administered to the subjects as part of a much larger project:

(a) The AH 4 group test of general intelligence (Heim, 1970). This test is composed of two parts, the first being concerned with verbal and numerical reasoning and the second with diagrammatical or spatial reasoning. The scores on each part as well as the total score were recorded for this study.

(b) The Manifest Anxiety Scale (Taylor, 1953).

(c) The scale for measuring symptomatic depression developed by Hathaway and McKinley (1963) from the original M.M.P.I. battery.

(d) The Extraversion Scale derived from the M.M.P.I. by Giedt and Downing (1961).

In addition to this subjects were each evaluated by the experimenter in terms of the Gunn Criminal Profile (Gunn and Robertson, 1976). With this procedure the subject receives a grading on a five-point scale in each of the following categories of criminal activity: theft, fraud, sexual offences, violent offences, motoring offences, drinking offences, drugs offences and offences resulting in financial gains. The experimenter takes into account his impressions of the prisoner after an intensive interview, together with police and prison records, in allocating the scores. In this way it is hoped to overcome the fact that official records do not necessarily reflect actual behaviour and to obtain a more realistic assessment of criminality.

Each of the subjects completed a grid by evaluating a set of nine predominantly authority figures on ten supplied constructs (see Table 13.1). They were administered in a semantic differential/type format with each of the elements being evaluated on all the constructs in turn. The grids were subjected to a principal component analysis with the INGRID program and the three structural measures extracted. Product-moment correlations were then calculated between the three measures and all the other measures including age and length of time spent in prison.

Part 3 One hundred and seventy-five male prisoners, age range 18—36, were the subjects of this experiment. As part of a much larger project the following tests were administered:

(a) The Progressive Matrices Test of Intelligence (Raven, 1956).

(b) A 110-item questionnaire (P.Q.) incorporating measures of neuroticism, extraversion and psychotism, and a lie scale (Eysenck and Eysenck, 1972).

TABLE 13.1   The elements and constructs included in
the grids for part 2

Elements	Constructs
A Social workers	1 Helpful/not helpful
B Prison officers	2 Insincere/sincere
C Father	3 Hardworking/lazy
D Magistrates	4 Dishonest/honest
E Governors	5 Considerate/inconsiderate
F Psychiatrists	6 Narrow-minded/broad-minded
G Myself	7 Friendly/unfriendly
H Prison doctors	8 Inhumane/humane
I Police	9 Good/bad
	10 Rude/polite

(c) The E.P.I. (Eysenck and Eysenck, 1964) to give measures of neuroticism and extraversion, and a lie score.

In addition, each subject completed a grid by evaluating six female physiques on ten constructs using a seven-point scale in a semantic differential format. (The elements and constructs are the same as those used in part 1 of this study.) The grids were analysed by the INGRID program, and measures of cognitive complexity, bias and variability extracted from the analysis. Intercorrelations were then calculated between all measures.

Table 13.2 gives the means and standard deviations of all the variables included in all the studies. Table 13.3 shows the correlations between the three structural measures and the other variables, together with significance levels.

Results

Part 1   The only correlation to reach significance concerning cognitive complexity was that with the lie scale: the higher the score on the lie scale, the more cognitively simple the subject. Among the response error measures, the only correlation to reach significance was that between neuroticism and variability: the higher neuroticism scorers showed a greater tendency to use extreme scores. Approaching significance was the negative relationship between age and variability: the younger subjects tended to exhibit more of the extreme scores than the older. There were also three sizeable relationships involving the bias score, but again not quite significant: high scorers on the psychotism scale tended to show high bias scores and low scorers on the sociability scale of the extraversion scale tended to show high bias scores.

Part 2   Four of the variables included in this study were significantly correlated with cognitive complexity. The older the subject and the longer the time spent in prison, the more cognitively complex he tended to be. The more offences he had committed involving fraud, the more likely he was to be cognitively complex, and, inversely, the more violent offences he had committed, the more likely he was to be cognitively simple. With regard to the response error measures, variability was

TABLE 13.2   Means and standard errors of all
variables included in parts 1, 2 and 3

Part 1 (N = 40)

Variable	Mean	S.e.
Cognitive complexity	55.13	1.52
Bias	0.45	0.02
Variability	0.55	0.02
Age	20.79	0.33
Psychotism	3.92	0.39
Sociability	7.95	0.47
Impulsivity	4.45	0.24
Extraversion	12.39	0.59
Neuroticism	10.42	0.82
Lie score	3.82	0.50

Part 2 (N = 95)

Variable	Mean	S.e.
Cognitive complexity	60.66	1.40
Bias	0.36	0.02
Variability	0.59	0.02
Age	27.17	0.58
Months in prison	50.19	4.17
Intelligence part 1	31.64	1.01
Intelligence part 2	40.11	1.18
Intelligence total	71.69	2.03
Depression	74.56	1.62
Manifest Anxiety Scale	82.85	1.74
Extraversion	46.04	1.08
Theft offences	2.85	1.00
Fraud offences	1.20	0.14
Sex offences	0.30	0.09
Violent offences	2.16	0.15
Motoring offences	1.75	0.13
Drinking offences	1.07	0.11
Drugs offences	1.17	0.14
Offences for financial gains	1.56	0.11

Part 3 (N = 175)

Variable		Mean	S.e.
	Cognitive complexity	52.98	0.771
	Bias	0.53	0.012
	Variability	0.61	0.012
	Age	26.35	0.269
	Intelligence	42.87	0.536
P.Q.	Psychotism	6.59	0.293
	Exraversion	14.30	0.388
	Neuroticism	12.80	0.406
	Lie score	5.81	0.296
E.P.I.	Neuroticism	26.88	0.775
	Extraversion	28.33	0.565
	Lie score	3.85	0.200

TABLE 13.3   Correlations between the three structural measures and the other variables

Part 1

Variable	Correlation with cognitive complexity	Correlation with bias	Correlation with variability
Age	−0.17	0.09	−0.27
Psychotism	−0.10	0.26	0.16
Sociability	−0.27	−0.25	−0.01
Impulsivity	0.06	−0.08	0.10
Extraversion	−0.19	−0.23	0.03
Neuroticism	0.14	−0.08	0.34*
Lie score	−0.34*	0.04	0.12

Part 2

Variable	Correlation with cognitive complexity	Correlation with bias	Correlation with variability
Age	0.21*	0.10	−0.10
Months in prison	0.22	0.03	0.00
Intelligence part 1	0.13	−0.12	−0.06
Intelligence part 2	0.17	−0.12	−0.08
Intelligence total	0.16	−0.13	−0.07
Depression	−0.03	−0.12	−0.01
Manifest Anxiety Scale	−0.16	−0.26*	0.12
Extraversion	−0.03	−0.03	0.21*
Theft offences	0.05	−0.27*	0.19
Fraud offences	0.24*	−0.04	−0.03
Sex offences	0.02	0.03	−0.04
Violent offences	−0.22*	0.05	0.07
Motoring offences	−0.17	−0.14	0.06
Drinking offences	−0.11	−0.05	0.19
Drugs offences	−0.14	−0.21*	0.14
Offences for financial gains	0.04	−0.06	0.19

Part 3

Variable		Correlation with cognitive complexity	Correlation with bias	Correlation with variability
	Age	0.05	0.07	−0.08
	Intelligence	−0.14	−0.13	−0.11
P.Q.	Psychotism	−0.01	0.00	−0.05
	Extraversion	−0.02	−0.01	−0.08
	Neuroticism	0.05	−0.01	0.23*
	Lie score	−0.01	−0.01	0.21*
E.P.I.	Neuroticism	0.13	0.03	0.23*
	Extraversion	0.07	0.03	−0.10
	Lie score	0.00	0.03	0.14

*Indicates $p < 0.05$.

significantly related to extraversion, indicating that the high extraversion scorers were also those who tended to use the extreme score positions more frequently. There was a significant negative correlation between bias and manifest anxiety, indicating that low anxiety subjects tended to be high bias scorers. There were also significant negative correlations between bias scores and theft and drugs offences, indicating that the prisoners exhibiting low bias tended to be those who were convicted on drug or theft charges.

Part 3 There were no significant correlations concerning the cognitive complexity measure. The neuroticism scale of both the P.Q. and E.P.I. was significantly correlated with the variability measure, indicating that high neuroticism scorers tended to be extreme responders. The lie scale of the P.Q. was also significantly correlated with the variability measure and the lie scale of the E.P.I. showed a positive relationship, but not a significant one. High liars therefore tended to be extreme responders.

Discussion

(*a*) *Cognitive complexity* The results concerning the correlation between cognitive complexity and personality variables as measured by the standard tests are interesting in their lack of support for such a relationship. In all studies none of the measures of psychotism, neuroticism, depression or extraversion were shown to be related to the measure of cognitive complexity. Most of the past research findings which have shown relationships of significance have used psychiatric patients as subjects. These findings suggest, however, that in a non-patient sample the relationships do not hold.

On the other hand, the aspects of personality displayed in the lie scale do show some relationships with cognitive complexity. In the normal sample of study 1, the subjects who scored high on the lie scale were characterized by more cognitively simple cognitive structures than those who scored low. The lie scale is intended to measure 'faking' tendencies in the respondent. So the individuals who showed a greater propensity to fake in their responses to the questionnaires were the more cognitively simple. This can perhaps be explained in the greater number of response options open to the cognitively complex and therefore the less opportunity to fake, whereas the cognitively simple person who has greater experience in situations where a faking response is possible is more likely to develop a set towards faking. This relationship was not displayed by the prisoner sample of study 3.

Further differences between the samples were exhibited by the relationship between age and cognitive complexity. No relationship of significance was evident in studies 1 and 3, whereas there was a direct correlation between the two measures in study 2. This, however, is unlikely to represent differences between the samples, but rather it is necessary to have a wide range of ages present, as in the study 2 sample, to find a relationship between age and cognitive complexity. A relatively small range of ages was represented in the other two studies. In the criminal sample of study 2, not only was cognitive complexity correlated with age but also with time spent in prison. So the older subjects tended to be the more cognitively complex, and those who had spent more time in prison also tended to be more complex. There is, of course, a high correlation between age and the length of time in prison, so it seems that one cannot conclude that spending time in prison tends to produce a greater degree of cognitive complexity, but rather the increase in age concurrent with time spent in prison does so. The logical opposite to these findings, however, is worthy of note. It suggests that spending time in prison does not affect the development of cognitive complexity and, therefore, that the increase in experiential level which accompanies increasing age in normal subjects is not prevented when the individual is imprisoned.

The relationships between cognitive complexity and types of criminal activity demonstrated in study 2 are also of interest. Only two of the categories of behaviour were related to the measure of cognitive complexity: the more cognitively complex subjects tended to have been involved in criminal activities involving fraud, whereas the more cognitively simple subjects tended to have committed offences involving violence. Interpretation of these findings seems very straightforward. To commit a fraudulent offence the subject needs to be able to conceptualize and plan the intricacies of the situation and to display a certain amount of creativity and imagination. All these qualities would appear more likely in the cognitively complex subject than the cognitively simple, whereas offences involving violence are more likely to be impulsive, uncalculated events. The more unidimensional person is likely to commit such acts in situations where violence seems to him the immediate course of action, whereas the more complex person is more likely to consider other options and therefore less often becomes involved in violent offences. This finding concerning the violent prisoner is in line with that of Topçu (1976), who showed that more aggressive subjects also tended to be more cognitively simple.

These findings on the whole provide much food for thought and bases for further study. If cognitive complexity is to be shown to be a personality trait, then a great deal more research into its correlates will have to be done. In the meantime these studies have provided some evidence both for and against the hypothesis.

(b) Response error measures   The overall picture of the personality correlates of the two response style measures indicates one predominantly consistent result: i.e. the relationship between neuroticism and variability. With all three measures of neuroticism included in the studies there was a positive, significant correlation with variability. This would imply that a subject exhibiting a high extreme response score would tend to be characterized by a neurotic personality, whereas a low variability scorer would tend not to exhibit neurotic traits. This finding is in line with previous research reported and provides further evidence of the relationship. The correlation can perhaps be explained in terms of the increased rigidity displayed by the more neurotic personality and hence the tendency to see things in definite black or white terms with little desire for the intervening shades – an attitude which would understandably result in frequent use of the extreme grades.

The positive correlation between lie scores and variability scores was significant in one of the three cases and of reasonable size in the other two. This indicates that subjects who tend to use the extreme response scores also have a tendency to 'fake' or dissimilate in the questionnaire situation. This is an interesting finding, indicating that those individuals who do tend to give untruthful responses do so in a confident manner. In terms of the interpretation of the variability score, therefore, it means that the experimenter should be slightly more wary of the high extreme responder's grid than that of the low. It perhaps includes more socially desirable responses, more responses to please the examiner, etc., and should be interpreted with allowances for these facts.

Only one of the four measures of extraversion displayed a significant relationship with variability. The other three showed near-zero positive and negative correlations. It hardly seems feasible, therefore, to attach much importance to the solitary relationship with the M.M.P.I. measure of extraversion other than as further evidence that the three Eysenckian measures (P.I., P.Q., E.P.I.) are assessing a different attribute from the American measure. The controversy concerning the comparability of personality tests cannot be discussed here, but the relationship obtained between extraversion and variability must be considered very tentative.

The relationship between intelligence and extreme responding reported by some writers was non-existent in these data. Four different intelligence measures were used but none of them approached a significant correlation with the variability measure. It seems therefore that the variability score cannot be interpreted in terms of intelligence.

Age was not found to affect variability scores, though these samples did not contain a wide enough range to replicate the previous findings of high extreme responding in young children and older adults. All the correlations were negative and one approached significance in the direction of decreasing variability with age.

The most important finding concerning the bias score was its significant negative correlation with manifest anxiety. This implies that those individuals who tend to place the elements to an equal extent on either side of the midpoint of the scale, and thus obtain low bias scores, are likely to be high on the Manifest Anxiety Scale. Alternatively, individuals who exhibited a high degree of bias are likely to be low on manifest anxiety. This phenomenon can be explained in terms of the ego-defence mechanisms employed by the anxious person. In order to overcome his anxiety the individual may develop certain rituals or compulsions. For example, the need for balanced and symmetric states is a common defence among anxious people. Hence in assigning grades to the elements in a grid by attempting to balance them on either side of the midpoint, the highly anxious individual is relieving some of his stress. On the other hand, those low on anxiety feel no such compulsion for equality and balance, and in fact are more relaxed and less activated people. Hence they tend to discriminate less between the elements and perhaps see them all in a more favourable light, and by so doing obtain a higher bias score.

The significance of this in terms of the interpretation of the bias score is that it indicates that a certain degree of bias is to be found in the 'normal' person and that a low bias score implies an abnormal person in that he is likely to be highly anxious.

The correlations between bias and the measures of psychotism and the sociability scale of extraversion, though not significant, were quite sizeable: high bias scorers also tended to be high psychotism scorers. This relationship seems reasonable because the psychotic person, as he becomes less in touch with the real world, utilizes fewer cues in evaluating other people and so becomes less discriminatory between them, resulting in a tendency towards a bias score. The negative relationship between sociability and bias indicates that the less sociable, more introverted person also tends to be less discriminatory. Again it could be hypothesized that this is because the introverted person is not as interested in other people as is the extrovert and hence uses fewer cues to evaluate them.

Other findings of importance concern the relationship between bias and intellectual functioning. Although all four measures of I.Q. were negatively related to bias to some degree, none of the correlations approached significance. Research findings in the past have shown a negative relationship between acquiescence and various cognitive variables, and the lack of significance in the correlations reported here is perhaps surprising. Similarly there was no relationship of importance between the measures of impulsivity and bias. Previous research has indicated a significant relationship between these two also. The explanation for these discrepant findings may be that the bias score is not measuring a variable that can be equated to acquiescence as defined in past research. Alternatively, it may further prove that the existence of such a phenomenon is suspect because of the lack of such confirmatory results.

A significant negative relationship was found between the bias scores and offences involving theft and drugs, so that theft and drug charge prisoners tended to be low bias scorers. This finding is difficult to interpret in its own right and would need more exploration to be meaningfully understood.

13.3 Conclusions

It is hoped that these findings have provided some evidence concerning the concomitants of the measures of cognitive complexity, bias and variability obtained with the INGRID program. We are still left with a great deal of uncertainty about the exact nature of the variables and can only conclude that a great deal more work needs to be done in this area. No doubt if the use of grid technique expands and increases at its present rate more data will become available and more light will be shed upon the nature and meaning of these measures. In the meantime the studies reported here may have provided some insights into the meanings of the variables and some guidelines for the direction of future research.

References

Arthur, A. Z. (1966). 'Response bias in the semantic differential'. *Br. J. soc. clin. Psychol.*, 5, 103–107.

Bannister, D. (1960). 'Conceptual structure in thought-disordered schizophrenics'. *J. Ment. Sci.*, 106, 1230–1249.

Bannister, D. (1962). 'The nature and measurement of schizophrenic thought-disorder'. *J. Ment. Sci.*, 108, 825–842.

Bannister, D. (1963). 'A genesis of schizophrenic thought-disorder: a serial invalidation hypothesis'. *Br. J. Psychiat.*, 109, 680–686.

Bannister, D. (1965). 'Genesis of schizophrenic thought-disorder: retest of the serial invalidation hypothesis'. *Br. J. Psychiat.*, 111, 377–382.

Bannister, D., and Mair, J. M. M. (1968). *The Evaluation of Personal Constructs*, Academic Press. London and New York.

Barnes, E. H. (1955). 'The relationship of biased test responses to psychopathology'. *J. abn. soc. Psychol.*, 51, 286–290.

Berg, I., and Collier, J. S. (1953). 'Personality and group differences in extreme response sets'. *Educ. psychol. Measmt.*, 13, 164–169.

Bieri, J. (1955). 'Cognitive complexity-simplicity and predictive behaviour'. *J. abn. soc. Psychol.*, 51, 263–268.

Bieri, J. (1961). 'Complexity-simplicity as a personality variable in cognitive and preferential behaviour'. In D. W. Fiske and S. Maddi (Eds.), *Functions of Varied Experience*, John Wiley and Sons, New York.

Bieri, J., and Blacker, E. (1956). 'The generality of cognitive complexity in the perception of people and inkblots'. *J. abn. soc. Psychol.*, 53, 112–117.

Bieri, J., and Messerley, S. (1957). 'Differences in perceptual and cognitive behaviour as a function of experience type'. *J. consult. Psychol.*, 21, 217–221.

Binner, P. R. (1958). *Permeability and Complexity: Two Dimensions of Cognitive Structure and Their Relationship to Behaviour*, Unpublished Ph. D. thesis, University of Colorado.

Blumenthal, R. (1968). 'Extreme response in the attitudes of schizophrenics'. *Psychol. Rep.*, 22, 586.

Bopp, J. A. (1955). *A Quantitative Semantic Analysis of Word Associations in Schizophrenia*, Unpublished Ph. D. thesis, University of Illinois.

Borgatta, E. F., and Glass, D. C. (1961). 'Personality concomitants of extreme response set (ERS)'. *J. soc. Psychol.*, 55 213–221.

Brengelmann, J. C. (1958). 'The effects of exposure time in immediate recall on abnormal and questionnaire criteria for personality'. *J. Ment. Sci.*, 104, 665–680.

Brengelmann, J. C. (1959). 'Differences in questionnaire responses between English and German nationals'. *Acta Psychologica*, 16, 339–355.

Brengelmann, J. C. (1960a). 'Extreme response set, drive level and abnormality in questionnaire rigidity'. *J. Ment. Sci.*, 106, 171–186.

Brengelmann, J. C. (1960b). 'A note on questionnaire rigidity and extreme response set'. *J. Ment. Sci.*, 106, 187–192.

Brown, E. G. (1964). *The Effects of Content Ambiguity on Response Sets in Two Populations*, Unpublished Master's thesis, University of Richmond.

Campbell, V. N. (1960). *Assumed Similarity, Perceived Sociometric Balance, and Social Influence*, Unpublished Ph. D. thesis, University of Colorado.

Cattell, R. B. (1955). 'Psychiatric screening of flying personnel'. *Personality Structure in Objective Tests — A Study of 1,000 Air Force Students in Basic Pilot Training*. Project No. 21-0202-0007, Report No. 9, Air University, USAF School of Aviation Medicine, Texas.

Cattell, R. B., Dubrin, S. S., and Saunders, D. R. (1954). 'Verification of hypothesized factors in one hundred and fifteen objective test designs'. *Psychometrika*, 19, 209–230.

Cattell, R. B., and Gruen, W. (1955). 'The primary personality factors in 11-year-old children, by objective tests'. *J. Pers.*, 23, 460–478.

Chetwynd, S. Jane (1975). *Register of Grid Users*, St. George's Hospital, London S.W.17.

Chetwynd, S. Jane (1976). 'Differences between men and women in their stereotyping of the roles of 'wife' and 'mother'. *Explorations*, Chap. 9.

Chetwynd, S. Jane, and Slater, P. (1976). *Differences between Quasis and Experimental Grids in Measures of Bias and Variability*, Unpublished study, University of London.

Couch, A., and Keniston, K. (1960). 'Yeasayers and neasayers: agreeing response set as a personality variable'. *J. abnorm. soc. Psychol.*, 60, 151–174.

Crockett, W. H. (1965). 'Cognitive complexity and impression formation'. In B. A. Maher (Ed.), *Progress in Experimental Personality Research*, Vol. 2. Academic Press, New York.

Dresser, I. G. (1969). *Repertory Grid Technique in the Assessment of Psychotherapy*, Unpublished M. Phil. thesis, University of London.

Elliott, L. L. (1961). 'Effects of item construction and respondent aptitude on response acquiescence'. *Educ. psychol. Measmt.*, 21, 405–415.

Eysenck, H. J., and Eysenck, S. B. G. (1964). *Manual of the Eysenck Personality Inventory*, San Diego Industrial and Educational Testing Service.

Eysenck, H. J., and Eysenck, S. B. G. (1968a). 'A factorial study of psychoticism as a dimension of personality'. *Multivariate Behavioral Research* (Special Issue), **1968**, 15–31.

Eysenck, S. B. G., and Eysenck, H. J. (1968b). 'The measurement of psychoticism: a study of factor stability and reliability'. *Br. J. soc. clin. Psychol*, 7, 286–294.

Eysenck, S. B. G., and Eysenck, H. J. (1971). 'Crime and personality: item analysis of questionnaire responses'. *Br. J. Criminology*, 11, 49–62.

Eysenck, S. B. G., and Eysenck, H. J. (1972). 'The questionnaire measurement of psychoticism'. *Psychol. Med.*, 2, 50–55.

Forehand, G. A. (1962). 'Relationships among response sets and cognitive behaviours'. *Educ. psychol. Measmt.*, 22, 287–302.

Frith, C. D., and Lillie, F. J. (1972). 'Why does the repertory grid test indicate thought-disorder?'. *Br. J. soc. clin. Psychol.*, 11, 73–78.

Gault, Una (1971). *'Relationships Between Some Structural and Content Measures of Cognitive Complexity*, Unpublished manuscript.

Giedt, F. H., and Downing, L. (1961). 'An extraversion scale for the M.M.P.I.'. *J. clin. Psychol.*, 17, 156–159.

Gunn, J., and Robertson, G. (1976). 'A criminal profile'. *Br. J. Criminology*, (In press).

Hall, M. F. (1966). *The Generality of Cognitive Complexity–Simplicity*, Ph. D. Thesis, Vanderbilt University.

Hamilton, D. L. (1968). 'Personality attributes associated with extreme response style'. *Psychol. Bull.*, 69, 3, 192–203.

Hathaway, S. R., and McKinley, J. C. (1963). In G. S. Walsh and W. G. Dahlstrom (Eds.) *Basic Readings on the M.M.P.I. in Psychology and Medicine*, Minnesota Press.

Haynes, E. T., and Phillips, J. P. N. (1972). *Inconsistency, Loose Construing and Schizophrenic Thought-disorder*, Paper read at the 1972 London Conference of the B. P. S.

Heim, A. W. (1970). *Manual of the AH 4 Group Test of General Intelligence*, National Foundation for Educational Research Publishing Co., Windsor, England.

Hesterley, S. O. (1963). 'Deviant response patterns as a function of chronological age'. *J. consult. Psychol.*, 27, 210–214.

Higgins, J. C. (1959). *Cognitive Complexity and Probability Preferences*, Unpublished manuscript, University of Chicago.

Hunt, D. E. (1962). *Personality Patterns in Adolescent Boys* Progress Report No. 7, Grant No. M-3517, Syracuse University.

Iwawaki, S., and Zax, M. (1969). 'Personality dimensions and extreme response tendencies'. *Psychol. Rep.*, 25, 31–34.

Jackson, D. N., and Pacine, L. (1961). 'Response styles and academic achievement'. *Educ. psychol. Measmt.*, 21, 1015–1028.

Jaspars, J. M. F. (1966). *On Social Perception*, Unpublished Ph. D. thesis, University of Leiden.

Jones, A. (1956). 'Distributions of traits in current Q-sort methodology'. *J. abnorm. soc. Psychol.*, 53, 90–95.

Jones, R. E. (1954). *Identification in Terms of Personal Constructs*, Unpublished Ph. D. thesis, Ohio State University.

Kelly, G. A. (1955). *The Psychology of Personal Constructs*, Vols. 1 and 2. W. W. Norton and Co., New York.

Kerrick, J. S. (1956). 'The effects of manifest anxiety and I.Q. on discrimination'. *J. abnorm. soc. Psychol.*, 52, 136–138.

Koch, E. (1958). *A Study of Conceptual Behaviour with Social and Non-social Stimuli*, Unpublished Ph. D. thesis, University of North Carolina.

Leventhal, H. (1957). 'Cognitive processes and interpersonal predictions'. *J. abnorm. soc. Psychol.*, 55, 176—180.

Lewis, N. A., and Taylor, J. A. (1955). 'Anxiety and extreme response preferences'. *Educ. psychol. Measmt.*, 15, 111—116.

Light, C. S., Zax, M., and Gardiner, D. H. (1965). 'Relationship of age, sex and intelligence level to extreme response style'. *J. Pers. and soc. Psychol.*, 2, 907—909.

Lo Guidice, A. (1963). 'Cognitive variables, trait centrality and conflict resolution in impression formation'. *Dissertation Abs.*, 23.

Love, H. G., and Priest, H. F. (1967). 'Extreme responding in psychiatric patients'. *N.Z. med. J.*, 66, 882—883.

Lundy, R. N., and Berkowitz, L. (1957). 'Cognitive complexity and assimilative projection in attitude change'. *J. abnorm. soc. Psychol.*, 55, 34—37.

Luria, Z. (1959). 'A semantic analysis of a normal and a neurotic therapy group'. *J. abnorm. soc. Psychol.*, 58, 216—220.

McGee, R. K. (1962). 'Response style as a personality variable: by what criterion?'. *Psychol. Bull.*, 59, 284—295.

Mayo, C. W. (1959). *Cognitive Complexity and Conflict Resolution in Impression Formation*, Unpublished Ph. D. thesis, Clark University.

Mayo, C. W., and Crockett, W. H. (1964). 'Cognitive complexity and primacy-recency effects in impression formation'. *J. abnorm. soc. Psychol.*, 68, 335—338.

Messick, S. J., and Frederiksen, N. (1958). 'Ability, acquiescence and 'authoritarianism'. *Psychol. Reps.*, 4, 687—697.

Messick, S. J., and Kogan, N. (1967). 'Categorizing styles and cognitive structure'. *Educational Testing Service Research Bull*, Princeton, N.J.

Mogar, R. E. (1960). 'Three versions of the F scale and performance on the semantic differential'. *J. abnorm. soc. Psychol.*, 60, 262—265.

Neuringer, C. (1961). 'Dichotomous evaluations in suicidal individuals'. *J. consult. Psychol.*, 25, 445—449.

Neuringer, C. (1963). 'Effect of intellectual level and neuropsychiatric status on the diversity of intensity of semantic differential ratings'. *J. consult. Psychol.*, 27, 280.

Nidorf, L. J. (1961). *Individual Differences in Impression Formation*, Unpublished Ph. D. thesis, Clark University.

Norman, R. P. (1969). 'Extreme response tendency as a function of emotional adjustment and stimulus ambiguity'. *J. consult. clin. Psychol.*, 33, 406—410.

O'Donovan, D. (1965). 'Rating extremity: pathology or meaningfulness'. *Psychol. Rev.*, 72, (5), 358—372.

Osgood, C. E., Suci, G. J., and Tannenbaum, P. H. (1957). *The Measurement of Meaning*, University of Illinois Press, Urbana.

Parsonson, B. S. (1969). 'Extreme response tendencies on the semantic differential'. *Psychol. Rep.*, 24, 571—574.

Peak, H., Muney, B., and Clay, M. (1960). 'Opposites, structures, defenses and attitudes'. *Psychol. Mono.*, 74, (8 whole No.495).

Pederson, F. A. (1958). *A Consistency Study of the R.C.R.T.*, Unpublished Master's thesis, Ohio State University.

Raven, J. C. (1956). *The Standard Progressive Matrices*, Lewis and Co., London.

Rosenkrantz, P. S. (1961). *Relationship of Some Conditions of Presentation and Cognitive Differentiation to Impression Formation*, Unpublished Ph. D. thesis, Clark University.

Runkel, P. J., and Damrin, D. E. (1961). 'Effects of training and anxiety upon teacher's preference for information about students'. *J. educ. Psychol.* , **52**, 254—261.

Scott, W. A. (1963). 'Cognitive complexity and cognitive balance'. *Sociometry*, **26**, 66—74.

Sechrest, L. B., and Jackson, D. N. (1961). 'Social intelligence and accuracy of interpersonal predictions'. *J. Pers.*, **29**, 167—181.

Shaw, M. E. (1961). 'Some correlates of social acquiescence'. *J. soc. Psychol.*, **55**, 133—141.

Signell, K. (1966). 'Cognitive complexity in person perception and nation perception: a developmental approach'. *J. Pers.*, **34**, 517—537.

Soueif, M. I. (1958), 'Extreme response set as a measure of intolerance of ambiguity'. *Br. J. Psychol.*, **49**, 329—334.

Stricker, G., and Zax, M. (1966). 'Intelligence and semantic differential discriminability'. *Psychol. Rep.*, **18**, 775—778.

Stringer, P. F. (1971). 'Cognitive structure, education and change'. In M. L. J. Abercrombie, P. F. Stringer and P. M. Terry, *Changes in Students' Performance on Personality Tests during University Courses*, A final report to the S.S.R.C. (unpublished).

Supnick, J. (1964). Cited in Crockett, W. H. (1965). 'Cognitive complexity and impression formation'. In B. A. Maher (Ed.), *Progress in Experimental Personality Research*, Vol. 2. Academic Press, New York.

Taylor, J. A. (1953). 'A personality scale of manifest anxiety'. *J. abnorm. soc. Psychol.*, **48**, 285—290.

Topçu, S. (1976). *Psychological Concomitants of Aggressive Feelings and Behaviour*, Unpublished Ph. D. thesis, University of London.

Tripodi, T., and Bieri, J. (1963). 'Cognitive complexity as a function of own and provided constructs'. *Psychol. Rep.*, **13**, 26.

Tripodi, T., and Bieri, J. (1964). 'Information transmission in clinical judgements as a function of stimulus dimensionality and cognitive complexity'. *J. Pers.*, **32**, 119—137.

Vannoy, J. S. (1965). 'Generality of cognitive complexity-simplicity as a personality construct'. *J. pers. soc. Psychol.*, **2**, 385—396.

Warren, N. (1964). *An Investigation of the Cognitive Structures of Two Social Class Groups*, Paper presented at the Annual Conference of the B.P.S.

Wertheimer, R., and McKinney, F. (1952). 'A case history blank as a projective technique'. *J. consult. Psychol.*, **16**, 49—60.

White, B. J., and Harvey, O. J. (1965). 'Effects of personality and own stand on judgement and production of statements about a central issue'. *J. exper. soc. Psychol.*, 1 334—347.

Wilkinson, A. E. (1970). 'Relationship between measures of intellectual functioning and extreme response style'. *J. soc. Psychol.*, **81**, 271—272.

Zax, M., Gardiner, D. H., and Lowy, D. G. (1964). 'Extreme response tendency as a function of emotional adjustment'. *J. abnorm. soc. Psychol.* , **69**, 654—657.

Zax, M., Loiselle, R. H., and Karras, A. (1960). 'Stimulus characteristics of Rorschach inkblots as perceived by a schizophrenic sample'. *J. proj. Tech.*, **24**, 439—443.

Zuckerman, M., and Norton, J. (1961). 'Response set and content factor in the Californis F scale and the Parental Attitude Research Instrument'. *J. soc. Psychol.*, **53**, 199—210.

Zuckerman, M., and Norton, J., and Sprague, D. S. (1958). 'Acquiescence and extreme sets and their role in tests of authoritarianism and parental attitudes'. *Psych. Res. Reps.*, **10**, 28—45.

14

GENERALIZED PERSONAL QUESTIONNAIRE TECHNIQUES

J .P.N. Phillips

Psychology Department, University of Hull

14.1 Introduction *(a) The problem*

Psychiatrists and clinical psychologists are confronted by patients, one of the features of whose illness is that they express a number of complaints, such as 'I feel depressed', 'Everything around me seems real and I'm not', 'I believe that "they" use microphones and electronic devices to keep a record of my every move', 'I can't concentrate', 'I believe that my sins will find me out', 'My mind is unclear', 'I feel I cannot go into the garden by myself', 'My thoughts go too fast for me to catch on to them', etc., etc.

Although such symptom statements usually constitute both a major factor in referral to treatment and an important index of response to it, they have been surprisingly little investigated. From them arise two questions of both theoretical (scientific) and practical (medical) importance. Firstly, how are fluctuations in them related to fluctuations in other indicators of the patient's illness, such as behavioural or physiological measures? Secondly, how do they respond to attempts to modify them experimentally or to cure the illness? To answer these questions, some measure or scaling of alterations in the level of the symptoms is needed. The most obvious approach would be to ask the patient directly about such changes. However, everyday language does not make possible any great precision of response. The patient may, for example, answer that the symptom has completely, or almost, or nearly completely disappeared, or that it has become very slightly, slightly, a little, a bit, rather, somewhat, perhaps, definitely, markedly, decidely, much or very much better or worse, or that it has not changed, or hardly, barely, or scarcely changed at all. Such replies do not readily lend themselves to any quantification which would be of scientific use. The problem is therefore to find some more precise and numerical scaling of fluctuations in the patient's symptoms.

(b) A solution

The solution to this problem offered by generalized personal questionnaire techniques is to construct, for an individual patient, a special form of questionnaire.

Every item which is derived from a statement of the type illustrated above, made by the patient and expressing a symptom of his illness, provides a scaling of the levels assumed by this symptom. Administrations of the questionnaire on a number of successive occasions will thus give a scaling of the successive levels of all the symptoms occurring in it.

Although clearly foreshadowed by the work of Stephenson (1953), Kelly (1955) and Osgood, Suci and Tannenbaum (1957), the device of constructing a questionnaire from statements made by an individual patient appears to have been first employed by Shapiro and Ravenette (1959) in an experiment on paranoid delusions; however, the form of the questionnaire used was conventional. The first true type of personal questionnaire was described by Shapiro (1961a, 1961b) and used mainly with female depressive patients. It was subsequently pointed out (Phillips, 1963) that Shapiro's form was a very special case and that there existed a general class of such techniques, of which Shapiro's was just one member. It is the purpose of this chapter to give an account of this class. However, Shapiro's original technique must first be described.

(c) Shapiro's personal questionnaire technique

Construction   In a standard interview, statements describing his symptoms are elicited from the patient. A questionnaire item is constructed from each statement in three steps.

Step 1   From each symptom statement (or statement 3, or illness statement, e.g. 'I feel *very* tensed up') the psychologist derives two further statements, one representing a somewhat lesser degree of intensity (statement 2, or improvement statement, e.g. 'I feel *rather* tensed up') and the other representing remission of it (statement 1, or recovery statement, e.g. 'I do *not* feel *at all* tensed up').

Step2   (The 'scaling procedure')   All the statements thus obtained are presented in a random order to the patient, who is asked to say which of the nine phrases shown in Table 14.1 (modified from the affective rating scale of Singer and Young, 1941) would best describe the state of mind of a person making the statement. Any statement not assigned to its desired position on the scale is immediately reformulated, in discussion with the patient, until he does so assign it.

Step 3   From each set of three statements three cards are prepared, each bearing a different pair of statements, one above the other, as illustrated in Table 14.2. The set of three cards constitutes a questionnaire item, and the constructed questionnaire consists of all the items.

Administrations   The questionnaire is administrated to the patient whenever desired (either for clinical reasons or as part of a therapeutic experiment) by presenting to him every card in turn, in a random order, and asking him to say which of the two statements on it comes closer to describing how he feels.

Scoring   Responses to each item are entered in the upper half of a 3 by 3 table in which, for any two distinct integers j and k between 1 and 3, a 1 in row j column k

TABLE 14.1   The modified Singer–Young affective
rating scale, with desired positions of statements

Very great unpleasantness
Very unpleasant   } (Illness) statement 3

Moderately unpleasant
Slightly unpleasant   } (Improvement) statement 2

Don't know

Slightly pleasant
Moderately pleasant   } (Recovery) statement 1
Very pleasant

Very great pleasure

indicates that statement k is preferred to statement j and a 0 indicates the reverse. The eight possible patterns of response to a single item are shown in Table 14.3. There are two different types of response pattern: inconsistent and consistent.

Inconsistent patterns   In pattern (V) and pattern (VIII) both statement 1 and statement 3 are preferred to statement 2: in other words, the patient has indicated that both the Illness and the Recovery statements describe his current position better than the intermediate Improvement statement, so that the pattern is inconsistent.

In pattern (VI) statement 1 is preferred to statement 2, which is preferred to statement 3, which is preferred to statement 1, while in pattern (VII) precisely the reverse occurs: since these two patterns each involve a circularity, they are also inconsistent.

Consistent patterns   The remaining patterns, (I), (II), (III) and (IV), are consistent, and represent different and increasing degrees of intensity of the symptom. They may be assigned scores 0, 1, 2 and 3 respectively, corresponding to the number of 1's they contain. (Shaprio equivalently scores them 1, 2, 3 and 4.)

Example 1   Table 14.4 shows the results from the first administration, every succeeding tenth administration and the last administration of a personal

TABLE 14.2   Illustration of
three cards constituting a Shapiro
personal questionnaire item

| I do *not* feel *at all* tensed up |
| I feel *rather* tensed up |

| I feel *very* tensed up |
| I feel *rather* tensed up |

| I feel *very* tensed up |
| I do *not* feel *at all* tensed up |

TABLE 14.3   Possible response pattern to an item in a Shapiro personal questionnaire, in the conventional representation

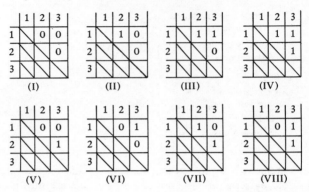

questionnaire which, except for its last item (which used four statements), was a Shapiro personal questionnaire. The following brief case history may illustrate the table.

The patient was referred by a general practitioner, and when first seen on 26 April, 1968 complained of a rather diffuse set of anxiety, phobic and depressive symptoms. The anxieties and phobias were treated by systematic desensitization, and over the first seventeen months there was slow but considerable improvement. About 18 September, 1968 the treatment seemed to reach a crisis in which no further progress was being made, some of the previously desensitized phobias began to reappear and the patient expressed the feeling that the unconscious fears were welling up to the surface but not quite getting there. Psychotherapy was discussed, but did not prove feasible. Instead, the desensitization procedure was modified in such a way as to ensure that the relaxation responses were overlearnt (see Phillips and Kenyon, 1975, where the patient is referred to as Mr. Y.). This expedient appeared to be successful, desensitization proceeded at an even more deliberate pace and his keeping of appointments, previously somewhat haphazard, became meticulous. For the last year of treatment the writer felt that desensitization was not achieving much, but the patient nevertheless seemed to find it beneficial. Finally, after gradually more widely spaced appointments, on 23 July, 1971 the patient appeared with the fully formed conclusion that this was the time to terminate treatment, which was accordingly done.

(d) Generalized personal questionnaire techniques

The extensions of Shapiro's technique described in this chapter stemmed from two sources. The first lay in certain miscellaneous, fairly practical questions which arose in actual use of the technique. For example, in constructing a questionnaire to quantify changes in the intensity of the delusions of a paranoid patient, it seemed appropriate to employ statements representing five different levels of certainty, namely absolute certainty that the belief was true, absolute certainty that it was false, complete uncertainty as to whether it was true or false, the feeling that it was probably true and the feeling that it was probably false. But with five and not three

TABLE 14.4 Shapiro personal questionnaire results from a patient undergoing treatment by systematic desensitization for diffuse phobias

	1968		1969				1970				1971	
	17/5	4/10	2/1	20/3	5/6	18/9	8/1	2/4	17/6	15/10	31/3	23/7
1 Tensed up	2	0	3	2	1	(2)	0	3	1	0	3	0
2 Apprehensive in a built-up area	0	2	1	1	1	1	2	0	0	0	0	0
3 Courage to do things	2	0	0	1	3	1	0	0	0	0	0	0
4 Trouble going into shops	2	3	2	2	1	1	1	0	0	0	0	1
5 Feeling everybody looking at me	(1)c	0	0	0	0	0	0	0	0	0	0	0
6 Uneasy with other people	2	0	2	0	0	0	0	0	0	0	0	0
7 Not feeling like doing old hobbies	3	0	3	2	2	1	0	2	3	0	2	0
8 A job to concentrate	3	1	3	3	1	2	2	2	2	0	0	0
9 Apprehensive about future	2	1	2	1	0	(2)	0	1	2	0	1	0
10 Depressed	(3)	0	0	2	3	2	1	0	2	0	0	1
11 Feeling not responsible for actions	1	0	0	1	(2)	(2)	0	0	0	0	0	0
12 Thinking not going to cope	1	0	1	1	1	2	1	1	0	0	0	0
13 Sleep badly	2	0	2	2	0	3	0	0	0	2	0	0
14 No interest in sex[a]	—	—	(3)	0	1	(3)	1	1	1	1	1	1
15 Tired[b]	—	—	—	—	—	—	2	3	2	0	2	0

[a]This item was added to the questionnaire on 11 October, 1968.
[b]This item, which, unlike the others, contained four statements (see below, Section 14.4), was added to the questionnaire on 9 October, 1969.
[c]Circled scores represent inconsistent responses (see below, Section 14.5).

statements about each symptom (as described by Shapiro), it was no longer immediately apparent which patterns of responses were consistent, or how to deal with those which were not. Another problem was raised by the relatively poor discriminative power of the technique: it provided only an ordinal scaling, indeed an ordinal scaling comprising only four points. The question thus arose as to whether it would be possible in some way to obtain greater precision. There was a number of such problems.

The second, and more important, source was theoretical. Despite its great elegance, there seemed to be something tantalizingly arbitrary about the actual form of the technique. The practical reasons which had led to this form were clear and evident, but from a theoretical point of view it appeared to have sprung mysteriously from nowhere, to be unrelated to any other questionnaire technique and to exist completely isolated in an otherwise empty theoretical space. It was therefore desirable to discover the missing theoretical framework and correctly locate the technique within it; only in this way could a satisfactory understanding be achieved.

The answer to the theoretical problem was found to lie in a certain definition of *scaling*.[1] Although different types of scales, e.g. ordinal, ordered metric, interval, ratio, etc., have been well known to, and studied by, psychologists for many years, this definition seemed to unify them in a way that was at least partially new and that offered certain advantages. For it was not simply a description of what a scaling is, but also a prescription of how to obtain a scaling of almost any set of objects whatsoever. Seen from the standpoint of this prescription, the personal questionnaire technique appeared as a very special kind of joint ordered metric scaling of both the possible levels of the symptom represented by the reference statement and the levels actually assumed by it on the various occasions of administration of the questionnaire. Now if this latter feature of the technique, that it consisted of a joint scaling of both the statements and the occasions, was taken as the essential one for the purpose of quantifying changes in the symptom, then it was apparent that this scaling need not necessarily be ordered metric, but could be ordinal, or interval, or of any other type; it was also evident that certain other features of the technique (discussed below) were not essential for this purpose. Thus it was possible to formulate the concept of a *generalized personal questionnaire technique,* defined as a particular kind of joint scaling of both a set of reference statements and a set of levels of a symptom on various occasions: moreover, corresponding to the different types of scale there would be different types of personal questionnaire technique, e.g. an ordinal type, an interval type, etc.

The consequences of this definition were fourfold. Firstly, it offered a complete solution to the theoretical problem: the original technique could be properly placed within a broader conceptual background and related to the psychological literature on scaling; its apparent uniqueness was explained by the fact that it was a special case, even among ordered metric questionnaires.[2]

Secondly, the solution to the general theoretical problem made possible a more unified approach to the various practical questions referred to above, to certain of which particular answers had already been suggested. Some of these answers could be accommodated within and extended by the general theory: e.g. the method which had been devised for dealing with the problem of inconsistent response patterns in an item of more that three statements was found to be a special case of an approach used by Slater (1960a, 1960b, 1961) for dealing with sets of inconsistent pair comparison preference judgements, and it was further found that this method applied equally to the case of an ordinal personal questionnaire technique (see below, Sections 14.3d and 14.4d). Other answers were found to have been superseded by the general theory: e.g. a tentative contribution to the problem of obtaining greater precision with the original technique became virtually obsolete when it was realized that the same aim could be achieved more efficiently by the use of a different type of technique, the interval type (*q.v.*).

Thirdly, by virtue of the fact that there was not one but a whole spectrum of types of personal questionnaire technique, the psychologist's stock of methods of scaling fluctuations in the symptomatology of an individual patient was greatly broadened. Since the different types had different advantages and disadvantages, one suitable to the needs of a particular investigation to be carried out could be more easily chosen. In a word, the whole approach was thereby rendered more flexible.

Fourthly, however, it had happened that the generalization of the original concept of a personal questionnaire technique had involved some modifications of it, and certain features which had been considered essential to it no longer appeared to be so in the wider context. These are discussed in detail below (Section 14.10b), but there is one difference which should be mentioned at the outset, to prevent the reader's confusion. It is that whereas the original technique was thought of as containing a preliminary scaling of statements and a testing procedure in which fluctuations in the symptom were measured, as two separate stages, in the more general approach these two stages are considered to be merely parts of a single scaling experiment (see Table 14.11 below).

(e) Plan of chapter

The account falls into two parts, the first of which introduces and explains the concept of a generalized personal questionnaire technique, whereas the second illustrates the concept with a number of examples. The first part begins by informally presenting and formally stating the definition of a scaling, referred to above, which leads immediately to a general psychological scaling method. It is then shown that the original personal questionnaire technique is a special case of this method, but one which contains a number of restrictions. By removing these restrictions, a generalized personal questionnaire technique can be defined as a scaling, by the general method, of a particular set of objects. In the second part of the account, the translation of this theoretical concept into practice is shown for several different types of scaling, each of which leads to a different type of generalized personal questionnaire technique.

14.2 The concept of a generalized personal questionnaire technique

(a) Scaling

Introduction   As a very rough first approximation, it may be said that scaling a set of objects with respect to a certain property consists of providing some numerical representation or specification of the degree to which each of them possesses it. There is, in fact, a number of different kinds of such representations or specifications, conveying different sorts and amounts of information, and each kind constitutes a particular type of scale.

This preliminary sketch for a definition may be illustrated by the following four scale types.

1, One of the simplest is the *ordinal*[3] type, which consists of ranking the objects with respect to the given property. A famous ordinal scale is the Mohs scale of hardness, which arranges a number of minerals in order of that property: other examples are given by preference orderings of objects, orderings of players or teams in competitions, candidates in examinations, etc. If ties are allowed, then the ordinal scaling is said to be *partial*[4], otherwise it is said to be *complete.* More formally, an ordinal scaling is described by the following definition.

Definition   An ordinal scaling of a set of objects specifies, for pairs of them, which member of the pair is greater. If this is specified for all pairs, then the scaling is complete; otherwise it is partial.

2. Slightly less simple and well known is the *ordered metric*:[5] this consists of an ordinal scale with additional information about the spacing of the objects; namely the absolute values (i.e. the numerical values, irrespective of sign) of differences between pairs of objects are ranked. There are no very well-known examples of ordered metric scaling, but the concept may be illustrated by the three games noughts and crosses, draughts and chess: chess is more difficult than draughts, which is more difficult than noughts and crosses, so there is a clear ordinal scaling of them with respect to difficulty, but it might also be said that the difference in difficulty between chess and draughts is less in absolute value than that between draughts and noughts and crosses, giving an ordered metric scaling. (Note here that it is not necessary to know the actual values of the differences in difficulty, or even to be able to specify how they might be determined, but only to be able to rank them. Thus an ordered metric scaling lies somewhere between an ordinal scaling and an interval scaling (see 3 below) with respect to the amount of information it provides.) As with ordinal scaling, an ordered metric scaling may be either *partial* (if some of the absolute differences in difficulty are tied) or *complete.* It is also worth noting that an ordinal scaling is always a partial ordered metric scaling, since, for example, the absolute difference between the first and the last objects must be greater than the difference between any other pair of objects: thus the absolute differences between objects can be partially ranked by the ordinal scaling, although not completely.

Since an ordered metric scaling is, in effect, an ordinal scaling of the values of the differences between pairs of objects, it is formally described as follows.

Definition   An ordered metric scaling of a set of objects specifies, for pairs of absolute differences (between pairs of them), which member of the pair is greater. If this is specified for all pairs, then the scaling is complete; otherwise it is partial.

3. Still more informative is the *interval* scale, which assigns numbers to objects or entities in such a way that only the zero point and unit are arbitrary. The various temperature scales are all interval scales (although some of them are ratio scales as well — see 4 below). They have different zero points — Kelvin (absolute) and Rankine at absolute zero, Centigrade and Réaumur at the freezing point of water and Fahrenheit at the lowest temperature obtainable with a mixture of ice and salt — and different units — the Fahrenheit unit being 1.8 times the Centigrade and Kelvin unit, which is 1.25 times the Réaumur unit — but are otherwise equivalent. Hence any one of these scales may be transformed into any other by an appropriate linear transformation, i.e. multiplication by a constant and addition of a constant: this property, that they are essentially unaltered by a linear transformation, is common to all interval scales (Stevens, 1946, 1951). They are therefore formally described as follows.

Definition An interval scaling of a set of objects specifies, for pairs of absolute differences (between pairs of them), their ratio.

4. Finally, the *ratio* scale is more informative even than an interval scale: it assigns numbers to objects or entities in such a way that only the unit is arbitrary. The Kelvin and Rankine scales of temperature, as well as being interval scales, are also ratio scales, since their zero points are at absolute zero. In other words, an object at a temperature of, say, $200°$ Kelvin (or Rankine) is in some sense twice as hot as an object at $100°$. Another common example of ratio scaling is given by scales of weight: grams and pounds are different units, but otherwise equivalent, so that either of them may be transformed into the other by an appropriate similarity transformation, i.e. multiplication by a constant, and this property, that they are essentially unaltered by a similarity transformation, is common to all ratio scales (Stevens, 1946, 1951), which are therefore formally described as follows.

Definition A ratio scaling of a set of objects specifies, for pairs of them, their ratio.

It is now possible to arrive at a general definition of scaling by induction from the above examples.

General definition of scaling Each of the four definitions above has the same form, stating that the particular scale specifies, for certain pairs of entities, some binary relation between the members of the pair. The general definition will therefore state that a scaling specifies a binary relation. This is illustrated in Table 14.5. Only one further consideration need be taken into account, namely that the two kinds of entities occurring in the four definitions are of different types. In two cases (ordinal and ratio scaling) the binary relation is between the objects themselves, whereas in the other two (ordered metric and interval scaling) the relation is between functions of them, namely between absolute differences between pairs of them. This disparity is easily removed by considering the relation to be always between functions of the objects: this immediately covers the two cases of ordered metric and interval scaling, and it can be made to cover the other two cases by the use of the identity function. The latter is a kind of dummy function (like addition of 0, or multiplication by 1) which leaves everything to which it is applied unchanged. Thus, in the cases of ordinal and ratio scaling, the relation is considered to be between identity functions of the objects. Now the general definition of scaling may be formally stated.

Definition   *A scaling of a set of objects specifies, for pairs of functions of them, a binary relation.*

This definition is illustrated, for the four types of scaling given above, in Table 14.5, but it is in fact perfectly general and holds for any type whatsoever. However, only these four will be considered in this chapter.

(b) General method of psychological scaling

The reader of one of the standard textbooks on psychological scaling, say Guildford's *Psychometric Methods* (1954) or Torgerson's *Theory and Methods of Scaling* (1958), is likely to be struck by the fact that the subject is in a most unsatisfactory state: there exists a profusion of largely unrelated techniques, but little or no unifying theoretical framework. A case can be made that this unfortunate state of affairs is a result of inadequacies in the techniques themselves. Many of them, particularly Fechnerian methods and methods stemming from Thurstone's 'equation of comparative judgement' or Torgerson's related 'equation of categorical judgement', would more properly be described as procedures for investigating mathematical models than as scaling techniques. That is, they are more concerned with setting up and testing some particular model of how the subject behaves in the scaling experiment than with how the objects being scaled appear to him. Many, including those just named, are also open to the fundamental theoretical objection of making arbitrary and in principle untestable assumptions: for example, Fechnerian scaling assumes that just noticeable differences represent equal subjective differences, but it could equally well be assumed that they represent equal subjective ratios (and, precisely, analogously with the equation of comparative judgement — Stevens, 1951, 1959): the two different assumptions give quite different scales (the second being a logarithmic transformation of the first), but there seems to be no way of deciding between them without having recourse to

TABLE 14.5   Illustration of the definition of scaling. The type of scaling in each cell specifies, for pairs of the entities (functions) at the left of its row, the binary relation at the head of its column. (Adapted from Phillips, 1963)

		Binary relation	
		Member which is greater	Ratio of the members
Entity (function)	Objects (identity)	Ordinal	Ratio
	Absolute difference between objects	Ordered metric	Interval

TABLE 14.6   Illustration of the general direct method of psychological scaling

To obtain	the subject may be presented with pairs of	and asked to judge
ordinal scaling	objects	which is greater
ordered metric scaling	pairs of objects	which pair has the greater absolute difference between its members
interval scaling	pairs of objects	the ratio of the absolute differences between the members
ratio scaling	objects	their ratio

the direct approach of asking the subject how the objects appear to him. Thus the scales obtained in this way depend as much on the mathematical model chosen as on the subject's judgements, and in fact sometimes completely flout the latter (Ament, 1900; Newman, 1933; Miller, 1947).

Many scaling techniques also suffer from practical disadvantages. Some, in which the subject is required to adjust a stimulus until it bears a specified relationship to another, fixed, one (stimulus production methods), require either a continuum or at least a farily large, finely graded set of stimuli: such techniques are therefore not applicable to the scaling of moderate sized or small sets of stimuli. Others, particularly Thurstonian techniques, require very large numbers of responses, and are therefore not applicable with a single, reasonably fatiguable subject. Fechnerian and Thurstonian techniques also have the disadvantage of breaking down completely when all the stimuli being scaled are perfectly discriminable from one another. (This suggests that there is something fundamentally wrong in their approach.) Now a general method of obtaining a scaling of a set of objects from a single subject, which is not open to any of the above objections or difficulties, follows immediately from the general definition of scaling given above. It consists simply of presenting appropriate subsets of objects to the subject and asking him to specify the relevant relations between them. This is spelt out in detail, for the four types of scale shown in Table 14.5, in Table 14.6.

Out of the many possible examples which could be provided to illustrate this method (including a number of short-cut techniques for dealing with large numbers of stimuli), just two, which are directly relevant to what is to follow, will be given.

Example 2 Ordinal scaling   A forty-year-old psychiatric patient[6] complained of anxiety in various situations. Nine such were elicited from him, and randomly numbered, as follows:

1. Sitting in the cinema
2. Shopping an empty shop
3. Going to a party (many strangers present)
4. Going to a party (people I know)
5. At a football match

206

TABLE 14.7

(a) Judgement

1 > 2
3 > 9
8 > 4
5 > 7
1 > 6
3 > 2
4 > 9
5 > 8
6 > 7
3 > 1
4 > 2
5 > 9
8 > 6
1 > 7
3 > 4
5 > 2
6 > 9
8 > 7
1 > 4
5 > 3
6 > 2
7 > 9
8 > 1
5 > 4
3 > 6
7 > 2
8 > 9
5 > 1
6 > 4
3 > 7
8 > 2
1 > 9
5 > 6
4 > 7
8 > 3
9 > 2

(b) Tabulation of the data of (a)

	1	2	3	4	5	6	7	8	9	Total
1	\	0	1	0	1	0	0	1	0	3
2	1	\	1	1	1	1	1	1	1	8
3	0	0	\	0	1	0	0	1	0	2
4	1	0	1	\	1	1	0	1	0	5
5	0	0	0	0	\	0	0	0	0	0
6	1	0	1	0	1	\	0	1	0	4
7	1	0	1	1	1	1	\	1	0	6
8	0	0	0	0	1	0	0	\	0	1
9	1	0	1	1	1	1	1	1	\	7

(c) Permutation of the data of (b)

	5	8	3	1	6	4	7	9	2	Total
5	\	0	0	0	0	0	0	0	0	0
8	1	\	0	0	0	0	0	0	0	1
3	1	1	\	0	0	0	0	0	0	2
1	1	1	1	\	0	0	0	0	0	3
6	1	1	1	1	\	0	0	0	0	4
4	1	1	1	1	1	\	0	0	0	5
7	1	1	1	1	1	1	\	0	0	6
9	1	1	1	1	1	1	1	\	0	7
2	1	1	1	1	1	1	1	1	\	8

6. On a bus seat, facing another person
7. Interview with the 'boss'
8. Buying something at the Post Office
9. Shopping – one other person in front of me

Each situation description was typed on the blank side of a 5 by 3 inch index card, and the cards were presented to him in pairs, with the request to say each time which situation caused him the greater anxiety. His responses are shown in Table 14.7(a), where, for example, '1 > 2' indicates that he judged that situation 1 caused him greater anxiety than situation 2.

Analysis of data   The results are, following Kendall and Smith (1939) and Kendall (1948), entered in a table such as Table 14.7(b), where for any two distinct integers j and k between 1 and n, the number of objects, a 1 in row j column k indicates that situation k was judged to cause more anxiety than situation j, and a 0 in row j column k the reverse.

The data are either (perfectly) consistent, i.e. there exists a unique ordering of the objects with which every one of the subject's judgements agrees, or they are inconsistent, i.e. for any ordering of the objects whatsoever there is at least one of the patient's judgements which disagrees with it.

If they are consistent, as is the case with the data of Table 14.7, then the row totals of the table will be $0, 1, 2, \ldots, n-1$ in some order, and the rows and columns can be simultaneously permuted in such a way that the rearranged table contains only 0's in the upper right-hand half and only 1's in the lower left-hand half, as shown in Table 14.7(c) for the data of Table 14.7(b) and the ordinal scaling of the objects is given by the permuted order of the rows (and columns), here 5, 8, 3, 1, 6, 4, 7, 9, 2.

If they are not consistent, as is, for example, the case with the artificial data of Table 14.8(a), then an estimate (how good this estimate is remains to be seen) of the original scaling is given by the 'nearest adjoining order' (Slater, 1960a, 1960b, 1961), i.e. the order with which as small a number as possible of the subject's judgements are inconsistent; this smallest number, i, serves as a measure of the degree of inconsistency in the data. Thus, for the data of Table 14.8, the nearest adjoining order, exhibited in (b), is 4, 6, 9, 5, 10, 3, 7, 2, 1, 8 and $i = 3$. (The actual determination of i and the nearest adjoining order(s) for such a pair-comparison experiment is a matter of some difficulty and is omitted here – see Slater, 1960b, 1961; Remage and Thompson, 1966; Phillips, 1967, 1969. However, as will be seen with the corresponding type of personal questionnaire the determination is perfectly simple.)

If the data are not consistent, then of course the ordinal scaling(s) given by the nearest adjoining order(s) cannot be accepted with as much confidence as if they were; indeed they may, if very inconsistent, be virtually worthless. A test of whether or not they are acceptable is given by the statistic i. If it is sufficiently small, then the possibility that the subject was effectively responding at random can be confidently rejected. Tables of significance for i for up to eight objects are given by Slater (1961); although they do not cover the case of Table 14.8, it can be calculated from formulae given by Slater (1961) that the probability of obtaining an i as small as, or smaller than, 3 for ten objects is 0.000456, so that in this (artificial) case the nearest adjoining order could be accepted.

Example 3 Interval scaling   In the course of constructing an interval-type personal questionnaire (*q.v.*) for a twenty-seven year-old male homosexual patient, five statements, representing different possible levels of intensity of his expressed fear of sexual intercourse with girls, were elicited from him and randomly numbered:

1. The idea of sexual intercourse with a girl makes me slightly anxious
2. The idea of sexual intercourse with a girl makes me frightened
3. The idea of sexual intercourse with a girl makes me slightly frightened
4. The idea of sexual intercourse with a girl makes me anxious
5. The idea of sexual intercourse with a girl does not make me anxious at all

It was desired to obtain an interval scaling of this set. The general direct method prescribes that all possible pairs of pairs of the statements should be presented to

TABLE 14.8

(a) Example (artificial) of an inconsistent
pair-comparison table

	1	2	3	4	5	6	7	8	9	10	Total
1	\	1	1	1	1	1	1	0	1	1	8
2	0	\	1	1	1	0	1	0	1	1	6
3	0	0	\	1	1	1	0	0	1	1	5
4	0	0	0	\	0	0	0	0	0	0	0
5	0	0	0	1	\	1	1	0	1	0	4
6	0	1	0	1	0	\	0	0	0	1	3
7	0	0	1	1	0	1	\	0	1	1	5
8	1	1	1	1	1	1	1	\	1	1	9
9	0	0	0	1	0	1	0	0	\	0	2
10	0	0	0	1	1	0	0	0	1	\	3

(b) Permutation of (a) into the nearest
adjoining order

	4	6	9	5	10	3	7	2	1	8	Total
4	\	0	0	0	0	0	0	0	0	0	0
6	1	\	0	0	1	0	0	1	0	0	3
9	1	1	\	0	0	0	0	0	0	0	2
5	1	1	1	\	0	0	1	0	0	0	4
10	1	0	1	1	\	0	0	0	0	0	3
3	1	1	1	1	1	\	0	0	0	0	5
7	1	1	1	0	1	1	\	0	0	0	5
2	1	0	1	1	1	1	1	\	0	0	6
1	1	1	1	1	1	1	1	1	\	0	8
8	1	1	1	1	1	1	1	1	1	\	9

the subject in turn, each time with the request to judge the ratio of their differences. However, it is usually simpler to employ the short-cut procedure of holding one standard pair of statements (with an arbitrary assigned difference) before the subject, while presenting him with every other pair in turn, and asking him to judge the difference between its members in terms of the difference between the standard pair.

In the present case statement 5 and another statement, not in the above set, 'The idea of sexual intercourse with a girl makes me very frightened', were chosen as the standard pair and typed at opposite ends of a strip of card, approximately 13 by 2½ inch, with the assigned difference, 100, clearly indicated between them. The strip was laid before the patient and then all possible pairs were presented to, and judged by, him. His judgements are shown in Table 14.9(a).

TABLE 14.9 (a) Pair-comparison judgements of the differences in severity of five statements by a psychiatric patient

Pair	Judged Greater	Difference
1 2	2	30
3 5	3	20
4 1	4	15
2 3	2	25
5 4	4	35
1 3	3	30
4 2	2	40
5 1	1	15
3 4	3	30
2 5	2	80

TABLE 14.9 (b) Tabulation of the data of (a)

	1	2	3	4	5	Total	Estimate
1		30	30	15	−15	60	−12
2	−30		−25	−40	−80	−175	35
3	−30	25		−30	−20	−55	11
4	−15	40	30		−35	20	−4
5	15	80	20	35		150	−30

TABLE 14.9 (c) The statements, in increasing order of severity, with their estimated interval scale values

Statement	Estimate
2 Frightened	35
3 Slightly frightened	11
4 Anxious	−4
1 Slightly anxious	−12
5 Not anxious at all	−30

TABLE 14.9 (d) The difference judgements which the patient would have had to produce in order to be perfectly consistent with the estimates

	1	2	3	4	5	Total
1		47	23	8	−18	60
2	−47		−24	−39	−65	−175
3	−23	24		−15	−41	−55
4	−8	39	15		−26	20
5	18	65	41	26		150

Analysis of data The judgements are entered in a square table, such as Table 14.9(b), having regard to the signs of the differences; the row totals are computed, divided by the number of objects and multiplied by −1, to give estimated interval scale values of the statements, as shown in Table 14.9(b) and (c).

Unlike ordinal scaling data, which can be, and often are, perfectly consistent, interval scaling data of this kind are virtually certain to contain some inconsistencies. That this is so for the present data is seen from Table 14.9(d), which shows the judgements that patient would have had to have made to be perfectly consistent with the estimated scale values. As with ordinal scaling, it is necessary to know whether or not the errors fall within permissible limits; it is, further, desirable to know also whether, irrespective of their magnitude, the errors show evidence of being in some way systematic (e.g. does the subject persistently overestimate large differences and underestimate small ones, or vice versa?). A test for both these questions is given by a special analysis of variance (Phillips, Hutchings and Thompson, 1967; see also Phillips, 1971b). shown in Table 14.10. The F ratio for estimates is highly significant, indicating that there is clear evidence of differences between the estimates for the various statements, whereas the F ratio for non-additivity is not significant, so that there is no evidence of systematic inconsistencies in the data.

(c) Shapiro's technique and the general method of scaling

In practice, as has been seen, Shapiro's technique consists of two stages: a construction stage and a stage of administrations and scoring. However, in principle each item in it can be considered as a single partial ordered metric scaling of a certain double set of objects: the double set consists of both the possible levels of intensity represented by the statements and also of the actual levels assumed by the symptom on the various occasions. This will now be shown.

Firstly, in the 'scaling procedure' of step 2 of the construction stage, it is required that the three statements be assigned to certain ordered regions of the modified Singer—Young scale (see Figure 14.1). Thus this 'scaling procedure' gives an ordinal scaling of the three statements along the dimension of hedonic tone. (It also provides certain other information, but this is not relevant to the present purpose and will not be discussed until later; see Section 14.10b). Further, it has already been seen that an ordinal scaling is always a partial ordered metric scaling; in the present instance, the ordinal scaling specifies that the absolute difference between statements 1 and 3 is greater than the absolute difference between either of

TABLE 14.10   Analysis of variance of the data of Table 14.9

Source	Degrees of freedom	Sum of squares	Variance estimate	Variance ratio, F	Corresponding probability, p
Estimates	4	12030.0000	3007.5000	11.05	<0.01
Non-additivity	1	9.5857	9.5857	<1	N.S.
Residual	5	1360.4143	272.0828		
Total	10	13400.0000			

them and statement 2, but it does not specify whether, for the particular patient, statement 2 is closer to statement 1 or to statement 3. Thus the 'scaling procedure' provides a partial ordered metric scaling of the three statements.

Secondly, the patient's judgements in the administrations stage of the questionnaire are clearly ordered metric in form. With each card presented to him he is required to say which of two absolute differences is greater, that between his present position and the upper statement or that between his present position and the lower statement. Note that here, also, the ordered metric scaling is incomplete; the patient is not asked to compare the absolute difference between his present position and one statement with the absolute difference between some pair of statements.

Each item of Shapiro's technique can thus be considered as a single partial ordered metric scaling of both the statements and the various positions of the symptom on different occasions. It is, however, subject to certain limitations and restrictions.

Firstly, it uses only three statements. This may be desirable, in order to keep down the size of the complete questionnaire, when there is a large number of items, although even then it might be advantageous to have more than three statements, and thus greater precision, for certain items dealing with symptoms which are clinically crucial. However, when there is not a large number of items, and thus in general, this restriction is unnecessary.

Secondly, it has been seen that both the 'scaling procedure' and the administrations stage of the questionnaire involve partial, and not complete, ordered metric scaling, i.e. there are certain pairs of absolute differences which the patient is not required to compare. Although it turns out that this restriction is desirable for certain practical reasons when the judgements are ordered metric in form (see below, Section 14.4), there is no reason why it should be so in general, and in fact it transpires that it is not.

Thirdly, however, there is another respect in which the scaling is partial. No judgements are required of the patient which involve direct comparisons of the level of the symptom on two or more different occasions. The restriction appears to be a desirable one, since such comparisons would almost certainly present the patient with too difficult a task of organizing his recollections.

Fourthly, the judgements are limited to being ordered metric in form, but there is no reason why this need be so in general. It will be shown that there are perfectly practicable types of personal questionnaire technique which involve other forms of judgements.

Fifthly, the scaling is along a dimension of hedonic tone. Although this dimension is the most appropriate one for depressive symptoms, others would seem to be more suitable for other types of complaint. For example, paranoid delusions would probably be better scaled along the dimension of certainty of truth — certainty of falsity. Thus, in general, any relevant dimension might be employed.

(d) Generalized Personal Questionnaire Techniques

From the above considerations, it is possible to arrive at the concept of a generalized version of Shapiro's technique, which will contain all the latter's essential elements but without undesirable or unnecessary limitations and restrictions. Such a generalization will consist of a single partial scaling of statements and levels of a symptom, and the scaling will be partial in that the levels on different occasions are not directly compared, but not necessarily in any other respect. Any convenient number of statements may be used, the scaling may be of any suitable type and any relevant dimension of scaling may be used. Thus, a generalized questionnaire technique may be defined as follows.

Definition   A generalized personal questionnaire technique is a partial scaling, by the general direct method, of a certain double set of objects, along some relevant dimension. The double set of objects consists, firstly, of the levels of intensity of a condition represented by a number of reference statements and, secondly, of the levels actually assumed by that condition on a number of different occasions. The scaling is partial in that no judgements directly comparing two or more different occasions are elicited.

TABLE 14.11   The theoretical structure of a generalized personal questionnaire technique. Comparisons involving both the entities at the head of the column and to the left of the row constitute the part of the questionnaire named in the cell. Thus, the preliminary scaling of statements involves comparisons of statements with statements: data such as those of Tables 14.7(b) or 14.9(b) would occur in this cell. Each administration of the questionnaire, on the other hand, involves comparisons of statements with the level of the symptom on that occasion: data such as those of Tables 14.12 or 14.17 would occur in this cell. Note that the two cells named 'administrations of questionnaire' are mirror images of one another. In the following tables, only the upper of these two cells will be used

	Statements	Occasions
Statements	Preliminary scaling of statements (sometimes optional)	Administrations of questionnaire
Occasions	Administrations of questionnaire	

In practice, a generalized personal questionnaire will consist, just as with Shapiro's technique, of two stages, a construction stage and a stage of administrations and scorings. The former will involve a preliminary scaling of statements, in which judgements of the appropriate form involving only the statements are elicited. The latter will involve a series of occasions on each of which judgements involving the statements and the level of the symptom on that occasion are obtained. All this is illustrated in Table 14.11, to which repeated reference is made below.

Just what is the appropriate form of judgement will depend upon the type of scaling employed. To every type of scale (ordinal, ordered metric, interval, ratio, etc.) there will correspond a type of personal questionnaire technique. The remainder of this chapter is devoted to the description of the three types of technique corresponding to ordinal, ordered metric and interval scales.

14.3 The ordinal personal questionnaire technique

(a) Introduction

In this technique (Phillips, 1970b) a preliminary ordinal scaling of the statements is obtained and then, on every occasion of administration, each statement is presented to the patient with the request to say whether he feels better or worse than the statement would indicate.

(b) Preliminary scaling of statements

This may be carried out by the method of *Example 2* above, or even by simply asking the patient to rank the statements.

Example 4   A middle-aged business man with a stammer of many years duration was referred for treatment by means of behaviour therapy. The following five statements, representing different levels of frequency of this symptom, were arrived at in discussion with him, randomly numbered, and consistently scaled by him in the order shown:

Stammering
3. During the past week I have been stammering all the time
4. During the past week I have been stammering almost all the time
1. During the past week I have been stammering some of the time
2. During the past week I have been stammering occasionally
5. During the past week I have hardly ever been stammering

In the course of treatment, two further symptoms emerged, namely an inability to stick up for himself, and tenseness, for which the following further sets of statements were obtained and ranked as follows:

Sticking up for self
2. During the past week I have hardly stuck up for myself at all
4. During the past week I have stuck up for myself occasionally
3. During the past week I have stuck up for myself fairly well
1. During the past week I have stuck up for myself most of the time
5. During the past week I have stuck up for myself perfectly well

Tenseness
2. During the past week I have been extremely tense
4. During the past week I have been very tense
5. During the past week I have been fairly tense
3. During the past week I have been slightly tense
1. During the past week I have not been tense at all

(c) Preparation of materials and administrations of the questionnaire

Each statement is typed on the blank side of a 5 by 3 inch index card. The questionnaire consists of all the cards thus prepared, arranged in random order, and is administered by presenting to the patient every card in turn and asking him to

TABLE 14.12 Responses of a psychiatric patient to three ordinal personal questionnaire items. For explanation, see text.

(a) Stammering

	April 13	20	27	May 4	11	18	25	June 1	8	29	July 6	13	20	27	Aug 3	10	17	31	Sept 20	28	Oct 5	18	26	Nov 1	8	15	22	29	Dec 6	24	Jan 7 (1968)	Feb 21 (1968)
3. All the time	0	0	0	0	0	0	0	0	0	0	1	0	0	1	1	0	0	0	1	1	0	1	1	1	1	0	0	1	1	0	0	1
4. Almost all the time	0	0	0	1	1	0	0	1	1	1	1	0	1	1	1	0	0	0	1	1	0	1	1	1	1	0	0	1	1	0	1	1
1. Some of the time	1	0	1	1	1	1	0	1	1	1	1	0	1	1	1	1	0	0	1	1	1	1	1	1	1	1	1	1	1	1	0	1
2. Occasionally	1	1	1	1	1	1	1	0	1	1	1	1	1	1	1	0	1	0	1	1	1	1	1	1	1	1	1	1	1	0	1	1
5. Hardly ever	1	1	1	1	1	1	1	1	1	1	1	1	1	1	1	1	1	1	1	1	1	1	1	1	1	1	1	1	1	1	1	1
Score	3	2	3	3	4	2	2 (1) (3)	4	4	5	4	5	3	3	3 (1) (3)	1	5	5	5	5	3	5	3	5	5	3	3	5	1	5	5	5

(b) Sticking up for self

	April 13	20	27	May 4	11	18	25	June 1	8	29	July 6	13	20	27	Aug 3	10	17	31	Sept 20	28	Oct 5	18	26	Nov 1	8	15	22	29	Dec 6	24	Jan 7 (1968)	Feb 21 (1968)
2. Hardly at all	0	0	0	0	0	0	0	0	0	0	0	0	0	0	0	1	0	1	0	0	0	0	1	0	0	1	1	1	1	0	0	0
4. Occasionally	0	0	0	1	0	0	0	0	0	0	0	0	0	0	0	0	0	0	0	0	0	0	0	0	0	0	0	0	0	0	0	0
3. Fairly well	1	1	1	1	1	0	0	1	1	0	0	0	0	1	0	0	0	0	0	0	0	0	0	0	0	0	0	0	0	0	1	0
1. Most of the time	1	1	1	0	1	1	1	0	0	0	0	0	1	0	0	0	0	0	0	0	0	0	0	0	0	0	0	0	0	0	1	1
5. Perfectly well	1	1	1	1	1	1	1	1	1	0	1	0	1	1	1	1	0	1	0	0	0	0	0	0	0	0	0	0	0	0	1	1
Score	3 (1) (3)	1	3 (1) (3)	1	0	0	1	1	0	0	1	1 (1)(1) (3)	1	0	0	0	0	0	0	0	0	0	0	0	0	0	0	0	0	0 (0)	0 (3)	2

(c) Tenseness

2. Extremely	0	0	0	0	0	0	0	0	0	0	0	0	0	0	0	0	1	1
4. Very	0	1	0	0	0	0	0	0	0	0	0	0	0	0	0	1	1	1
5. Fairly	1	1	0	1	1	0	0	1	1	1	1	1	1	1	1	0	1	1
3. Slightly	1	1	0	1	1	1	0	1	1	1	1	1	1	1	1	1	1	1
1. Not at all	1	1	1	1	1	1	1	1	1	1	1	1	0	1	1	1	1	1
Score	3	4	2	1	3	3	1	2	3	3	4	3	3	3 (0) (2)	3	3	2	5

say whether the present intensity of his symptom is greater or less than the level represented by the statement on the card, i.e. informally, whether he feels 'better or worse than it says on the card', or some similar paraphrase: for example, the patient mentioned above preferred to say whether the activity mentioned in the statement was 'more' or 'less' than it said on the card.

(d) Scoring

The patient's responses on each occasion are entered in successive columns of a table such as Table 14.12. Each block of rows represents a different item and each row of a block a different statement of that item. (The rows of the blocks are arranged in decreasing order of severity of the statements.) If, on a particular occasion, the patient said he felt worse than indicated by the statement, then a 1 is entered in the appropriate cell, otherwise a 0. (Compare Table 14.12 with Tables 14.7(c) and 14.11.)

The set of responses, or response pattern, to the statements of an item on a particular occasion will be either consistent or inconsistent. A response pattern is *consistent* if the patient says that he feels worse than a statement at a given level of severity and all milder ones, but better than the next most severe and all severer ones (or worse than all statements, or better than all statements), in which case the current position of his symptom lies somewhere between these two statements (or is worse than, or better than, all statements) and thus consists of a column of nought or more 0's above a column of nought or more 1's. Consistent patterns may conveniently be scored by the number of 1's they contain, as has been done in Table 14.12 (unbracketed numbers at the feet of columns).

Any other pattern must contain a 1 above a 0, so that the patient has indicated both that he feels better than one statement and also that he feels worse than a more severe one, and is thus inconsistent. Analogously to Slater's rationale for obtaining an ordinal scaling from an inconsistent pair-comparison table, as in Table 14.8, such inconsistent patterns may be scored by the 'nearest adjoining pattern' (or patterns), i.e. as if they were the consistent pattern with which as few as possible of the patient's judgements are inconsistent. However, whereas with an inconsistent pair-comparison table it is a matter of some difficulty to determine the nearest adjoining order, here the nearest adjoining patterns may rapidly and easily be determined by comparing the inconsistent pattern with each of the possible consistent ones in turn and noting those which differ from it least. This is illustrated in Table 14.12 by the patterns with a bracketed number or numbers below them.

Once obtained and scored, the patient's responses on each occasion may, if there is a small number of items, be graphed, as shown in Figure 14.1 for the data of Table 14.12, affording an ongoing chart of the patient's progress. (For this purpose, the scores of inconsistent patterns with more than one nearest adjoining order may be obtained by averaging.) The following case history may briefly illustrate the graphs.

The patient complained of stammering in a number of social situations, particularly while telephoning or speaking to his superiors at work; the feature common to all these situations appeared to be someone else listening to him speak. Preliminary training in relaxation was given and a hierarchy of stammer-inducing situations drawn up and scaled, and on 4 May, 1967 attempted systematic

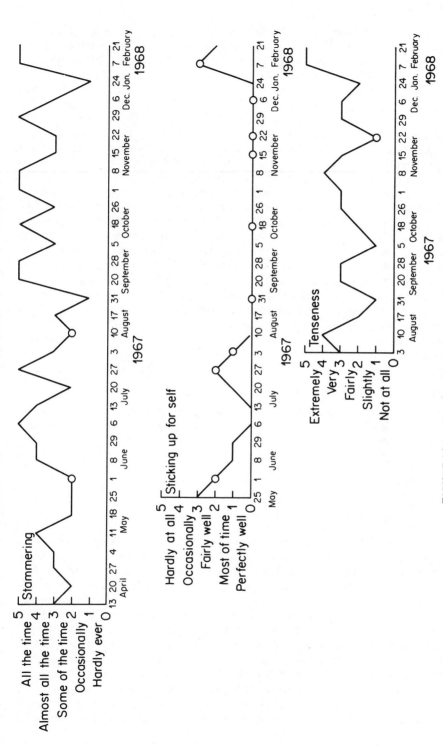

FIGURE 14.1   Graphs of the data of Table 14.12

desensitization of them began. On 11 May he also brought to light a difficulty in standing up for himself in social situations, which was fairly rapidly and effectively treated, in parallel with the less amenable stammering, by Wolpe's (1958) technique of 'reciprocal inhibition' by self-assertive responses. On 13 July the desensitization was supplemented by practice in 'shadowing' (Cherry and Sayers, 1956) and an item to scale his feelings of tenseness was included. By 31 August a considerable improvement in all three symptoms appeared to have been effected, but at the next session, when the writer returned from holiday, on 20 September, a marked relapse in stammering and tenseness was evident. Treatment continued in a rather unsatisfactory and ineffective way until 20 December, when the patient wrote to apologize for the temporary inability to attend for treatment, as a result of the death of an office superior and consequent increased work commitments, and also a slight heart attack of his own. He was not seen again until 24 January, 1968, when, in spite of the extra strain and further anxieties caused by an American takeover of his firm, he reported great improvement in stammering and tenseness. (This may have been because there was now no one in the office to listen to him telephoning.) However, the situation deteriorated rapidly, and on 21 February he recounted the apparent original traumatic cause of his stammering (which he recalled only dimly, but which had just been described to him by his brother), an incident in his early childhood when he had been shut up in the dark by a nursemaid. He requested abreaction treatment and was accordingly referred back to the responsible consultant psychiatrist, and behaviour therapy was terminated.

Inconsistent response patterns

Certain further points relating to inconsistent response patterns are discussed below (Section 14.5) after an account of ordered metric personal questionnaires has been given.

14.4 Ordered metric personal questionnaire techniques

(a) Rationale

It has already been seen that the types of judgement required of a patient in the administrations of a Shapiro personal questionnaire are ordered metric in form. Specifically, the patient is required to judge which is greater, the absolute difference (or distance) between the current level of his symptom and one statement, or the absolute difference between that level and another statement. These are not, however, the only possible judgements of ordered metric form which might be employed. For example, the patient might be required to compare the absolute difference between the current level and one statement with the absolute difference between that statement and another, or possibly even with the absolute difference between two other statements. But although the theory of techniques making use of such judgements has been worked out (Phillips, 1968c), and they even appear, in principle, to offer certain advantages such as greater precision and fuller coverage of the scale continuum, they would set the patient what seems to be

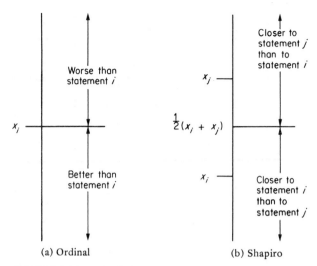

FIGURE 14.2   The judgements in an ordinal and a Shapiro item respectively. x_i and x_j are the scale values of statements i and j.

an extremely difficult and confusing task, and will therefore not be further discussed here.

Considering, then, only the type of judgement occurring in Shapiro's technique, it can be seen to be very similar in principle to that occurring in an ordinal questionnaire, as illustrated in Figure 14.2. In the latter the patient specifies whether the current level of his symptom is above or below the level represented by a statement, whereas in the former he specifies which of two statements it is closer to, i.e. whether it is above or below the midpoint, or average, of the pair of statements. (Compare Figure 14.2(a) with Figure 14.1.)

This similarity extends to an item in an ordinal questionnaire and an item in Shapiro's technique respectively, as illustrated in Figure 14.3. In each case, the judgements required of the patient determine a set of *boundary points* on the scale continuum (representing the levels of statements and the levels of midpoints of pairs of statements respectively in the two cases). A consistent pattern assigns the patient's current position uniquely to a region of the continuum, either above the highest point, or between a pair of adjacent points, or below the lowest point. The similarity between the two types of item can be brought out further by exhibiting the response patterns to a Shapiro item, shown in Table 14.3, in the same convention as for those of an ordinal item, as illustrated in Table 14.13.

Now in order to make use of this similarity, between an ordinal item of three statements and an item in Shapiro's technique, to determine the form of an item making use of any number of the type of judgement occurring in Shapiro's technique, it is necessary to determine the consistent patterns of such an item; and to determine the consistent patterns, it is necessary to determine the order of the boundary points of the corresponding regions. With an ordinal item this is no problem since, as has just been seen, the boundary points are simply the levels of the statements, which have already been ordered in the preliminary scaling of

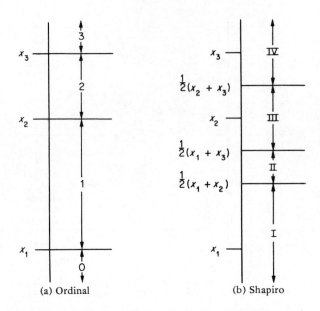

FIGURE 14.3   The regions defined by consistent response patterns in an ordinal and a Shapiro item respectively. x_1, x_2 and x_3 are the scale values of statements 1, 2 and 3.

statements. With an item making use of the type of judgement occurring in Shapiro's technique this is more difficult, since the boundary points are the midpoints of pairs of statements. With only three statements, as in Shapiro's technique itself, the solution happens to be very simple, because the ordering of three points necessarily orders their midpoints (see again Figure 14.3); however, with more than three statements a difficulty arises. This difficulty is that the ordering of four statements does not determine whether the midpoint of statements 1 and 4 is above or below the midpoint of statements 2 and 3; this will depend upon whether statements 1 and 2 are closer together or further apart than are statements 3 and 4, as illustrated in Figure 14.4. Moreover, with more than four statements there will be similar uncertainties associated with every other subset of four of them. This uncertainty thus makes it impossible, without further information, to determine whether some response patterns are consistent or not; e.g. if the patient judges that he is closer to statement 3 than to statement 2, and

TABLE 14.13   Possible response patterns to an item in a Shapiro personal questionnaire (from Table 14.3) shown in the convention for an ordinal item (see Table 14.12)

Pair	(I)	(II)	(III)	(IV)	(V)	(VI)	(VII)	(VIII)
2,3	0	0	0	1	1	0	1	1
1,3	0	0	1	1	0	1	0	1
1,2	0	1	1	1	0	0	1	0

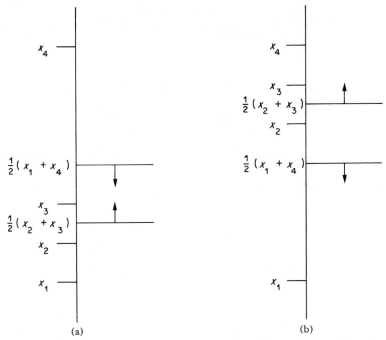

(a)   (b)

FIGURE 14.4   Illustration of the fact that the relative positions of certain midpoints depend upon the ordered metric scaling of the statements. If (a) statements 1 and 2 are closer together than statements 3 and 4, then the midpoint of statements 2 and 3 will be below that of statements 1 and 4, whilst if (b) they are further apart, then the reverse holds. Now suppose that the patient has judged that statement 3 better describes how he feels than statement 2, and statement 1 better than statement 4. Then in case (a) he has consistently indicated that the current level of the symptom lies between the two midpoints, whilst in case (b) he has inconsistently indicated that it is below the lower and above the upper

closer to statement 1 than to statement 4, as illustrated in Figure 14.4, then this pair of judgements will be consistent in one case and inconsistent in the other.

There are two ways out of this difficulty. The first, which is the natural way for an ordered metric item, is to determine the relative positions of all these uncertain pairs of midpoints by means of a preliminary ordered metric scaling of the statements. A full description of this method has been given elsewhere (Phillips, 1966): the details, which are slightly intricate (since it turns out that only a certain partial ordered metric scaling is necessary), will not be presented here, for the resultant technique tends to cluster the boundary points close together; as explained below (Section 14.6), this has the disadvantage that it would tend to require unduly fine discriminations of the patient and hence to give rise to an undesirably large number of inconsistent response patterns.

The second way, suggested by Shapiro (personal communication), is to omit from each item certain of the judgements required of the patient. This may be illustrated with the four-statement item mentioned above. If either (a) the patient is not asked to say whether statement 1 or statement 4 comes closer to describing

TABLE 14.14   Consistent response patterns
to the two possible four-statement ordered
metric items (shown in the same convention as
Table 14.13)

(a) Pair	0	1	2	3	4	5
3,4	0	0	0	0	0	1
2,4	0	0	0	0	1	1
2,3	0	0	0	1	1	1
1,3	0	0	1	1	1	1
1,2	0	1	1	1	1	1

(b) Pair	0	1	2	3	4	5
3,4	0	0	0	0	0	1
2,4	0	0	0	0	1	1
1,4	0	0	0	1	1	1
1,3	0	0	1	1	1	1
1,2	0	1	1	1	1	1

how he feels or (b) he is not asked to discriminate similarly between statements 2 and 3, then the uncertainty of the relative positions of the two midpoints will not matter, and all consistent response patterns, which are shown in Table 14.14 (using the same convention as Table 14.13) for these two possible omissions, can be identified. In fact, they have the same form as consistent patterns for an ordinal item, namely a column of nought or more 0's above a column of nought or more 1's.

With more than four statements, more than one judgement must be omitted. This must be done in such a way that the midpoints, corresponding to those remaining, can be rank ordered. The method has been given elsewhere (Phillips, 1966, 1968a); it consists of requiring from the patient only those judgements whose corresponding cells (in the Shapiro convention for a response pattern, as shown in Table 14.3; e.g. the judgement involving statements 1 and 3 corresponds to the cell in row 1, column 3) lie on a path through the table which does not move either upward or to the left. This is illustrated in Figure 14.5 by the two possible four-statement items already shown in Table 14.14(a) and (b), and also by the five possible five-statement items (c), (d), (e), (f) and (g). For more than three statements, there is always more than one possible item; the psychologist is free to choose the one which he considers the most suitable, but the present writer prefers those which satisfy the additional conditions that all statements in the item should, as far as possible, occur an equal number of times in the judgements required of the patient (for others, see Ingham, 1965; Mulhall, 1967). A list of all items with up to eight statements satisfying this additional condition is given in Table 14.15.

After this introductory explanation of its theoretical basis, the ordered metric technique may now be described fairly briefly.

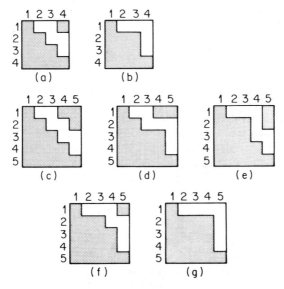

FIGURE 14.5   Illustration of the principle of construction of ordered metric items with more than three statements: the patient is only required to compare those pairs of statements whose corresponding cells lie on a path through the response pattern table which moves always downward and/or to the right. (a) and (b) are the two possible items with four statements, and (c) to (g) are the five possible ones with five statements. (a), (b) and (f) have the additional property that all statements occur, as far as possible, an equal number of times in the comparisons

(b) Preliminary scaling of statements

As with the ordinal technique, this may be carried out by the method of *Example 2*, or simply by asking the patient to rank the statements.

(c) Preparation of materials and administrations of the questionnaire

A set of pairs of statements is selected for each item. With a three-statement (Shapiro's) item, this set consists of all the possible pairs; with more than three statements, one of the sets in Table 14.15 may be used, or else any other set, fulfilling the condition of Figure 14.5, which the psychologist desires.

Once the set of pairs of statements for each item has been selected, then, as in Shapiro's technique, the questionnaire is constructed by typing every pair on a card, one statement above the other (which statement is above which below should be decided at random). The set of all the cards for all the items, arranged in random order, constitutes the questionnaire pack.

The questionnaire is administered, as in Shapiro's technique, by presenting each card to the patient in turn with the request to say which of the two statements on it comes closer to describing how he feels.

(d) Scoring

The patient's responses on each occasion are entered in a table analogous to that for the ordinal technique. As with that technique, a response pattern will be either consistent or inconsistent: consistent patterns consist of a column of nought or more 0's above a column of nought or more 1's; all other patterns are inconsistent. This is illustrated by Table 14.13, which shows all possible response patterns to a three-statement (Shapiro's) item, and where patterns (I) to (IV) are consistent whereas patterns (V) to (VIII) are inconsistent, and also by Table 14.14, which shows all the consistent patterns for either of the two possible four-statement items.

TABLE 14.15   Ordered metric items for four, five, six, seven and eight statements, with the additional property that all statements occur, as far as possible, an equal number of times in the pair comparisons

		Number of statements					
4		5	6	7	8		
A	B				A	B	C
Pairs	Pairs	Pairs	Pairs	Pairs	Pairs	Pairs	Pairs
3,4	3,4	4,5	5,6	6,7	7,8	7,8	7,8
2,4	2,4	3,5	4,6	5,7	6,8	6,8	6,8
2,3	1,4	2,5	3,6	4,7	5,8	5,8	5,8
1,3	1,3	2,4	3,5	4,6	5,7	5,7	4,8
1,2	1,2	1,4	2,5	3,6	4,7	4,7	4,7
		1,3	2,4	3,5	4,6	3,7	3,7
		1,2	1,4	2,5	3,6	3,6	3,6
			1,3	2,4	3,5	2,6	2,6
			1,2	1,4	2,5	2,5	2,5
				1,3	2,4	2,4	1,5
				1,2	1,4	1,4	1,4
					1,3	1,3	1,3
					1,2	1,2	1,2

As with the ordinal technique, consistent patterns may be scored by the number of 1's they contain and inconsistent ones by the nearest adjoining pattern, and again, once scored, the response patterns may be graphed to give an ongoing chart of the patient's progress, although it is not possible here, as it is with the ordinal technique, to enter the statements on the ordinate in any very simple and meaningful way.

Examples   An example of an ordered metric personal questionnaire has already been given in Example 1 above (see also Woodward and others, 1973).

14.5 Inconsistent responding in the ordinal and ordered metric techniques

(a) Introduction

The problems posed by inconsistent response patterns to ordinal or ordered metric items have been largely postponed above for consideration in the present section. They may be organized into the following logical sequence. To start with, it must be decided whether the data give evidence of sufficient consistency to warrant further consideration or whether they contain so much error that they can only be rejected as worthless. If the latter, then of course there is nothing more to be done with them, but otherwise two further problems arise. Firstly, some method must be found of estimating the consistent patterns which lie hidden behind this error: this has already been given above by the principle, following Slater's rationale for dealing with inconsistent pair-comparison preference tables, of scoring inconsistent patterns by the nearest adjoining consistent pattern. Secondly, it is desirable to attempt to diagnose the cause of the error and, if possible, to rectify it; two possible causes will be considered, an incorrect preliminary scaling of the statements and difficulties of discrimination on the part of the patient.

(b) Significance test for consistency

As with ordinal scaling, this is based on the statistic i, the number of discrepancies between a pattern and its nearest adjoining pattern(s). This is readily determined by comparing the obtained pattern with each of the possible consistent patterns in turn, and noting for each of them the number of responses which differ from those in the obtained pattern: the minimum number obtained is i and the pattern(s) with this minimum number of discrepancies is (are) the nearest adjoining pattern(s). (For example, with all consistent patterns $i = 0$, and with every one of the inconsistent patterns in Table 14.12 it so happens that $i = 1$.) The probability of different values of i under the null hypothesis of completely random responding has been obtained, in a slightly different context, by Slater (1960b), and is shown in Table 14.16(a). These probabilities may be used to carry out an exact test of whether Σi, the sum of all the i's, is less than would be expected on the basis of completely random responding. A simple computer program has been developed to carry out the computations (Phillips, 1970b). For the five-statement item 'Sticking up for self', with $\Sigma i = 9$ over twenty-six occasions, the exact probability is 0.000000007, so that the null hypothesis of no consistency at all may confidently be rejected. (However, it is evident that this item nevertheless has an unsatisfactorily large number of inconsistent response patterns, and possible causes of this high degree of error are considered in the following subsection.) An approximate test, suitable for hand computation, may be carried out by means of the calculated expected values and variances of i shown in Table 14.6(b) whenever the expected value is reasonably large. For the same item, the expected value of Σi is $26 \times 1.125 = 29.25$ with variance $26 \times 0.484375 = 12.59375$, so that

$$\chi^2 = \frac{(29.25 - 9)^2}{12.59375} = 32.56 \text{ with 1 d.f.}, p < 0.001$$

Either the approximate or the exact test (as appropriate) may also be carried out for all responses on a particular occasion.

TABLE 14.16

(a) The probability of Slater's i for an ordinal or ordered metric item
with three, four, five, six, seven or eight boundary points, i.e.
statements (ordinal) or statement pairs (ordered metric)

	$n = 3$	$n = 4$	$n = 5$	$n = 6$	$n = 7$	$n = 8$
$i = 0$	4/8	5/16	6/32	7/64	8/128	9/256
$i = 1$	4/8	9/16	16/32	25/64	36/128	49/256
	$i = 2$	2/16	10/32	27/64	56/128	100/256
		$i = 3$	5/64	28/128	84/256
			$i = 4$	14/256

(b) Expected values and variances
of i, calculated from (a)

n	$E(i)$	$V(i)$
3	0.50000	0.250000
4	0.81250	0.402344
5	1.12500	0.484375
6	1.46875	0.624023
7	1.81250	0.714844
8	2.17578	0.848007

(c) Incorrect preliminary scaling of statement

If the patient repeatedly produces inconsistent response patterns to a particular
item, the possibility that the preliminary ordinal scaling of the statements was
incorrect, either through arranging them in the wrong order or through failing to
reveal that they lie on more than one dimension, should be considered. In the
present case it is evident in retrospect that the statements for the item 'Sticking up
for self' were multidimensional, some of them (1, 2 and 4) referring to the
frequency and some (3 and 5) to the *quality* of this activity. An incorrect
preliminary scaling can be rectified by rescaling and, if necessary, reformulating the
statements, although this was (reprehensibly) not done here.

(d) Difficulties of discrimination

Suppose that, on a particular occasion of administration of a questionnaire, the
patient's position lies close to one of the boundary points of the regions
determining consistent response patterns, so that he has some difficulty in judging
whether it is above or below. Then one of two (or three) situations may arise.
Firstly, if there is no other boundary point close to the first and to the patient's
position, then the worst that can happen is that he may make one error of
discrimination, i.e. say he is above this boundary point when he is below it (or vice
versa), and thus produce a consistent response pattern which is adjacent to the one
which correctly represents how he feels. Secondly, if, however, the patient's
position lies between two boundary points which are close to one another, then he
may make two errors of discrimination and say that his position is above the upper

and below the lower, giving rise to an inconsistent pattern like those occurring in the items 'Stammering' and 'Tenseness'. (Thirdly, if several points lie close together, then more complicated errors could arise.) Thus, roughly speaking, difficulties of discrimination will at worst only give rise to slightly incorrect but still consistent response patterns *unless* two or more boundary points lie close together; unfortunately, it is not possible to be more precise. The remedy here again lies in reformulating the statements, or possibly omitting some.

14.6 Comparison of ordinal and ordered metric items

As has been seen, ordinal and ordered metric items are extremely similar, differing only in the type of judgements required of the patient. This difference has five consequences.

Firstly, the ordinal item is somewhat simpler and more natural, and consistent response patterns can be more directly related to the individual statements, as in Figure 14.1.

Secondly, the ordered metric item offers the possibility of greater precision of scaling with the same number of statements, since the number of boundary points is, for more than three statements, greater than the number of statements. This possible advantage is, however, counterbalanced by the possible disadvantage that the greater number of boundary points may lead to more inconsistent response patterns resulting from errors of discrimination. (This disadvantage would be intensified with the more complete type of ordered metric technique, not considered here, in which there is a preliminary ordered metric scaling of the statements and in which all possible pairs are presented to the patient.)

Thirdly, as is evident from Figure 14.3, with the ordered metric type the boundary points tend to cluster in the centre of the scale continuum, giving poor coverage to the extremes, so that fluctuations within the latter regions would not be identified as such.

Fourthly, the two types of judgement may be of different difficulties, although clinical experience to date with the two techniques does not suggest any very marked disparity.

Fifthly, the ordinal technique is open to a type of bias, namely a tendency on the part of the patient to say 'better' (or 'worse') irrespective of the content of the statements, to which the ordered metric is not. However, since the aim of all generalized questionnaire techniques is to scale changes in, and not the absolute level of, the patient's symptoms, this susceptibility is not such a weakness as might at first sight appear.

14.7 Schemes for automatic or rapid scoring of ordinal and ordered metric questionnaires

The scoring by hand of an ordinal or ordered metric personal questionnaire can be both tedious and lengthly. Several scoring schemes have been devised to avoid this waste of valuable skilled human time.

(a) Computer scoring

Phillips (1968b) reported a scheme for scoring Shapiro's three-statement ordered metric questionnaire, which was subsequently (Phillips, 1969, 1971a) extended to deal with a general ordinal or ordered metric questionnaire.

In the case of an ordinal questionnaire, each card of the questionnaire is a computer card; the statement is typewritten on the blank (reverse) side and the card is punched, in the first few columns, with the number of the item and the number of the statement. In an administration of the questionnaire, the patient sorts the cards into a 'better' pile and a 'worse' pile, according to whether he feels better or worse than the statement says. These two piles are headed by computer cards with the words 'BETTER' and 'WORSE' printed on their blank sides, and punched, in the first column, with the letters B and W respectively. After the administration the two piles, headed by their 'better' and 'worse' marker cards, are assembled together into a data pack (preceded by cards giving the patient's name, the number of items, etc., and followed by a terminator card) for computer analysis.

The case of an ordered metric questionnaire is exactly analogous, with the exceptions that two statements are typewritten on each card and the patient sorts the cards into a 'top' pile and a 'bottom' pile, according to whether the upper or the lower statement comes closer to describing how he feels.

I.C.T. 1900E ALGOL programs have been developed which check the data pack for obvious errors (such as missing or incorrectly punched cards) and score the questionnaire; this takes about five minutes.

A further feature of the scheme is that if random integers are punched in the last two columns of the cards, then a card sorter can be used to produce a (restrictedly) random ordering of the questionnaire pack, something that ordinary shuffling is rather inefficient at doing. If this is done, then the computer programs will also reconstruct the complete sequence of responses and carry out tests for several types of irrelevant response set: with an ordinal questionnaire, a tendency either to alternate or to persist with 'better' or 'worse' responses; with an ordered metric questionnaire, tendencies to prefer 'top' or 'bottom' responses, to alternate or persist with them and to alternate or persist with the 'better' or 'worse' responses they imply.

(b) The computer-assisted psychometric system (C.A.P.S.)

Sambrooks and MacCulloch (1970, 1973) and MacCulloch and Sambrooks (1974) reported a system using a desk top with a white reflecting surface upon which the statement pairs of a modified version of the S.O.M. (a standard questionnaire in the form of an ordered metric questionnaire with twelve five-statement items; see Feldman and others, 1966) were projected by means of a slide projector, and with two response buttons; this peripheral interface was connected to an encoder which recorded the subjects' responses and their latencies on paper tape for computer scoring. This system has been applied in the study of the effects of aversion therapy upon homosexuals.

(c) The personal questionnaire rapid scaling technique (P.Q.R.S.T.)

Mulhall (1976) has reported a technique which employs a standard set of eight phrases (absolutely none, almost none, very little, little, moderate, considerable, very considerable, maximum possible) to describe the intensity of each symptom, giving rise to eight-statement items or, by omission of the second and last phrases, six-statement items. Ingeniously devised printed answer sheets (on which each symptom is written by the user), reusable booklets and scoring stencils permit very rapid scoring. Mulhall has illustrated the routine use of this technique with six elegant clinical examples.

14.8 The interval personal questionnaire technique

(a) Introduction

In this technique (Phillips, 1970a) an (optional) preliminary interval scaling of the statements may be carried out and then, on every occasion of administration, each statement is presented to the patient with the request to judge the difference between his current position and the statement in terms of a standard difference.

(b) Preliminary scaling of statements

This may be carried out by the method of *Example 3*.

Example 5 A twenty-eight year-old man was referred for treatment by behaviour therapy of his desires for young boys. It also transpired on initial interview that the quality of normal sexual relations with his wife left something to be desired. Accordingly, the following two sets of five statements each, representing different levels of the two symptoms, were drawn up:

Having a rub with a boy
1. I think that having a rub with a boy would be very pleasant
2. I think that having a rub with a boy would be moderately pleasant
3. I think that having a rub with a boy would be indifferent
4. I think that having a rub with a boy would be slightly pleasant
5. I think that having a rub with a boy would give me very great pleasure

Having a go with my wife
1. Having a go with my wife gives me very great pleasure
2. Having a go with my wife is slightly pleasant
3. Having a go with my wife is very pleasant
4. Having a go with my wife is moderately pleasant
5. Having a go with my wife is indifferent

Since the five degrees of pleasure mentioned, which are taken from the Singer—Young affective rating scale (Singer and Young, 1941), have been found to be clearly distinct and discriminable to a number of subjects, no preliminary scaling of statements was carried out.

(c) Preparation of material and administrations of the questionnaire

For each item, a strip of card (about 13 by 2½ inch has been found to be convenient) is prepared with a standard pair of statements typed on it, one at each end, and a standard difference indicated between them. In the present case statements 3 and 5 were chosen as a standard pair for the first set and statements 5 and 1 for the second set, and the standard difference, arbitrarily chosen to be 100, marked in the centre of the strips. Also, each of the statements is typed on a separate 5 by 3 inch index card, the sets of cards being assembled into a separate pack for each item.

Each item of the questionnaire is administered by first laying the strip before the patient and, if it is thought necessary, reminding him of the difference between the members of the standard pair, and then presenting to him, in random order, each card of the pack in turn, with the request to say whether the present intensity of his symptom is greater or less than the level represented by the statement, and by how much in terms of the standard difference.

(d) Scoring

The responses on each occasion are entered in a table such as Table 14.17, where the convention has been adopted that responses indicating that the patient was more ill than the statement are taken to be positive and responses indicating that he was less ill negative. (Compare Table 14.17 with Tables 14.9(b) and 14.11.) Estimates of the intensity of the symptom on each occasion are given by the average of the judgements to each of the cards on that occasion, as illustrated in the table.

Once obtained, the patient's responses on the various occasions may be graphed as they are scored, as shown in Figures 14.6 and 14.7 for the data of Table 14.17, affording ongoing charts of the patient's progress. At the end of treatment or the experiment, estimates of the scale values of the statements may be obtained and entered on the ordinate, as shown in the figures, so that the different levels the symptom has assumed may be directly compared with those of the statements. Where, as here, there is no preliminary scaling of statements, the estimates for the statements are obtained by taking row totals, standardizing them by subtracting their mean from each, dividing by the number of occasions and changing signs, as illustrated in Table 14.17. The following case history may briefly illustrate Figures 14.6 and 14.7. (The year of treatment has been suppressed so as to make it less easy to identify the patient.)

At initial interviews on 3 May and 10 May the two interval items were constructed and possible therapeutic methods were discussed. It was decided to make a two-pronged attack on the patient's problem: attempts would be made both to increase the pleasantness of his relations with his wife and to decrease the pleasantness of his socially undesirable relations with boys. The former was initiated on 10 May by informal discussions and by lending him for study a book (Eichenlaub, 1964) giving instruction in normal sexual techniques; it was hoped that this would both dispel a certain ignorance and perfunctoriness in these matters which he displayed and also, by causing a mild degree of pleasant sexual arousal in association with imagined situations involving his wife, automatically induce an increased conditioning of sexual responses towards her. The latter would be implemented, as a last resort, by means of aversion therapy.

Over the period 17 May to 31 May he reported a gradual improvement in relations with his wife, but no change in his desire for boys. On 7 June he announced that he had been arrested and charged with indecently assaulting a male person under the age of sixteen years, the incident having taken place in the

TABLE 14.17   Responses of a psychiatric patient to two interval personal questionnaire items (for explanation, see text)

(a) Boy

	17/5	24/5	31/5	7/6	21/6	5/7	13/7	20/7	21/7	Total	Standardized Total	Estimate
5 Very great pleasure	0	-20	-10	-20	-15	-50	-75	-30	-100	-320	-239	26.5556
1 Very pleasant	20	-10	0	-20	-10	-20	-60	-30	-60	-190	-109	12.1111
2 Moderately pleasant	50	0	10	0	-10	-20	-25	0	-40	-35	46	-5.1111
4 Slightly pleasant	40	5	5	5	0	-10	-20	-10	-20	-5	76	-8.4444
3 Indifferent	100	0	20	20	5	0	0	0	0	145	226	-25.1111
Total	210	-25	25	-15	-30	-100	-180	-70	-220	-405		
Estimate	42	-5	5	-3	-6	-20	-36	-14	-44			

(b) Wife

	17/5	24/5	31/5	7/6	21/6	5/7	13/7	20/7	21/7	Total	Standardized Total	Estimate
5 Indifferent	-30	-30	-30	-20	-20	-100	-75	-50	-80	-435	-280	31.1111
2 Slightly pleasant	0	-20	-20	-15	-15	-50	-60	-30	-40	-250	-95	10.5556
4 Moderately pleasant	10	-10	-30	-10	-10	-30	-50	-10	-10	-150	5	-0.5556
3 Very pleasant	20	0	-10	0	-15	-20	0	0	0	-25	130	-14.4444
1 Very great pleasure	30	10	0	10	0	0	10	20	5	85	240	-26.6667
Total	30	-50	-90	-35	-60	-200	-175	-70	-125	-775		
Estimate	6	-10	-18	-7	-12	-40	-35	-14	-25			

Mr. C.   Having a rub with a boy

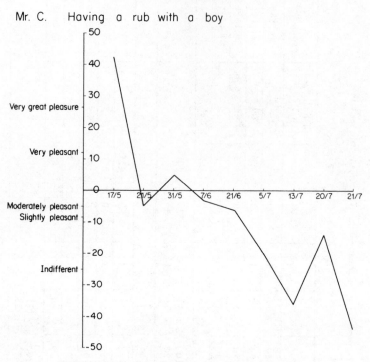

FIGURE 14.6   Graph of the data of Table 14.17(a)

Mr. C.   Having a go with my wife

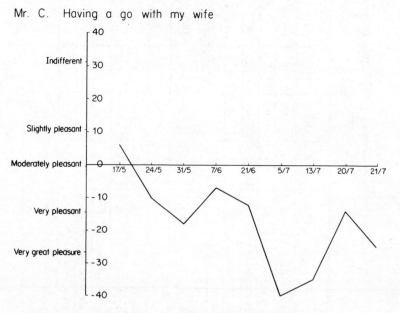

FIGURE 14.7   Graph of the data of Table 14.17(b)

preceding week. When the charge was heard by the Magistrates Court, he was put on probation on condition that he received aversion therapy. After some delay in preparing the necessary apparatus, he was admitted to hospital and aversion therapy began on 20 July: however, after one morning's treatment he refused to continue, against the most vigorous advice of the referring psychiatrist and the author, and was accordingly discharged.

(e) Final analysis of data

The same question of the acceptability of the data arises here as it does with the ordinal and ordered metric techniques, and is answered by an extension, shown in Table 14.18, of the analysis of variance for interval scaling which was illustrated in Table 14.10.

There are really three things which it is desirable to know. Firstly, are there any real changes in the symptom, or are the apparent changes accountable for in terms of random error? Secondly, do the statements represent satisfactorily distinct levels of intensity of the symptom? Thirdly, are the inevitable errors in the patient's judgements in any way systematic? These questions are answered by the F ratios for occasions, statements and non-additivity respectively (the F ratio for mean difference, although statistically necessary, is not here or usually psychologically meaningful), the first two of which are very highly significant in both items and the third of which is very highly significant for item (b), 'Having a go with my wife', but not significant for item (a). The significance of the F ratios for occasions and statements is desirable, since it indicates real effects, whilst that of the F ratio for non-additivity is not, since it indicates some degree of distortion in the patient's

TABLE 14.18   Analyses of variance of the data of Table 14.17

(a) Boy

Source	Degrees of freedom	Sum of squares	Variance estimate	Variance ratio, F	Corresponding probability, p
Occasions	8	24790.000	3098.7500	16.660	<0.001
Statements	4	14218.889	3554.7222	19.112	<0.001
Non-additivity	1	55.209	55.2092	0.297	N.S.
Mean difference	1	3645.000	3645.0000	19.597	<0.001
Residual	31	5765.902	185.9968		
Total	45	48475.000			

(b) Wife

Source	Degrees of freedom	Sum of squares	Variance estimate	Variance ratio, F	Corresponding probability, p
Occasions	8	8147.778	1018.472	7.643	<0.001
Statements	4	17994.444	4498.611	33.760	<0.001
Non-additivity	1	1904.695	1904.695	14.294	<0.001
Mean difference	1	13347.222	13347.222	100.164	<0.001
Residual	31	4130.861	133.254		
Total	45	45525.000			

Mr. C. Having a go with my wife

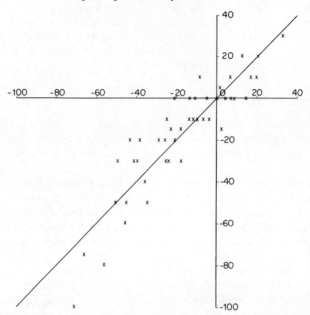

FIGURE 14.8   Plot of the judged differences of Table
14.17(b) against those predicted from the estimates for the
occasions and the statements. The abscissa shows the estimated
differences and the ordinate the judged differences

responses. A more detailed picture of this is given by Figure 14.8, which shows a
plot of the patient's responses against those which he should have produced
(compare Table 14.9d) to be perfectly consistent with the estimated values. It is
clear from this figure that the inconsistency lies in a tendency to exaggerate large
differences in either direction; in particular, the patient was overstating the
improvement in sexual relations with his wife.

14.9 Comparison of the interval technique with the ordinal and ordered metric
techniques

The interval technique differs from the ordinal and ordered metric techniques in
one minor and two major respects. The minor respect is that it makes use of a
standard difference defined by a pair of statements with an arbitrary value or
modulus. Standards of this kind are well known to have a slight distorting effect,
although the degree of this effect with an interval personal questionnaire, which
would be expected to depend both on the position of the statements and on the
modulus, remains a matter for future investigation. The two major respects are
different aspects of the fact that the interval technique uses quantitative
judgements of differences, whereas the ordinal and ordered metric techniques use
simply judgements of greater or less.

The advantage of quantitative judgements is that they permit much greater precision of scaling, namely an interval scaling rather than the four, five, six, etc. − point scaling obtained with judgements of greater or less. Moreover, since the interval judgements lead naturally to an analysis of variance rationale, this opens up further possibilities, which for lack of space can only be briefly indicated here, of testing hypotheses about the data. Thus, it is possible to test the values for the occasions when some treatment was applied against those when it was not, with considerably greater precision than for ordinal and interval techniques, since the standard error of estimate for each occasion value can be obtained from the analysis of variance. Again, the sum of squares for occasions can be partitioned into orthogonal polynomials due to trend, etc. (on both points, see Phillips, 1970a).

The converse of this advantage is that more difficult judgements are required of the patient, so that the interval technique may be less generally applicable than the other two. Thus, whereas Shapiro's technique, which is a special case of the ordered metric technique, has been successfully used with a patient with a Mill Hill vocabulary scale I.Q. as low as 83 (Shapiro, 1961c), the interval technique has so far only been tried with patients of above-average intelligence.

14.10 Conclusion

(a) Introduction

In this final section generalized personal questionnaire techniques are compared with Shapiro's original technique, the techniques are evaluated in the light of the traditional criteria of reliability and validity, possible applications of them are discussed and types of technique which have not been described here are briefly indicated.

(b) Comparison of generalized personal questionnaire techniques with Shapiro's original technique

Having given an outline account of generalized personal questionnaire techniques, it is now necessary to return to their starting point, Shapiro's original technique, and consider the differences between the two, and whether anything essential has been lost in the generalization.

Two major differences stand out: firstly, Shapiro's 'scaling procedure' has been replaced by a (sometimes optional) preliminary scaling method which does not have recourse to the modified Singer−Young affective rating scale and, secondly, the form of judgements required of the patient in administrations of the questionnaire varies with the type of questionnaire and is only the same in the case of ordered metric questionnaires. Each difference will be considered in turn.

Shapiro's 'scaling procedure' In the 'scaling procedure' just three statements are used, positioned at certain ordered regions on the dimensions of hedonic tone (see Table 14.1). These positions were chosen on the basis of their clinical relevance, it being considered desirable, *ceteris paribus*, to treat symptoms at the highest level, to continue treatment of symptoms at the intermediate level and cease treatment of symptoms at the lowest level. However, this feature of the scaling procedure does not appear to be indispensable. In the first place, it seems likely that decisions to treat, or to continue or cease treatment, would be taken on the basis more of clinical criteria than questionnaire responses. (After all, the patient might be attempting to deceive either the psychologist or himself.) Further, it is not possible with Shapiro's three-statement ordered metric item (although it would be with an ordinal or interval item) to tell whether a symptom has reached a level which is a least *slightly* pleasant; even if the patient produces response pattern (I) in Tables 14.3 and 14.13, this only indicates that his position is closer to *slightly* pleasant than to *moderately* unpleasant, which is not the same thing (see Figure 14.3b). Again, by the same token, there may be significant variation above response pattern (IV) in Tables 14.3 and 14.13 which is not measurable with this placement of the statements (see again Figure 14.3b). And finally, the psychologist may not consider the dimension of hedonic tone the most appropriate one along which to scale the symptoms of a particular patient, which may include delusions of persecution or grandeur, blunting of affect, or cognitive symptoms such as difficulties of concentration and memory disorders.

The form of judgements required of the patient   With Shapiro's original technique, the form of judgements required of the patient in administrations of the questionnaire was chosen so as to minimize the effects of irrelevant response set and to maximize reliability. This form of judgement is retained by ordered metric questionnaires, but not by ordinal or interval ones. It has been seen above (Section 14.6) that the ordinal item is open to a type of bias, a tendency to say 'better' (or 'worse') irrespective of the content of the statement being judged, of which the ordered metric item is free. The interval item is also open to this type of bias, and probably to more complicated ones involving numerical responses as well. However, as noted above (Section 14.6), a constant bias is unlikely to be a serious disadvantage in techniques for scaling changes.

As far as reliability is concerned, experience to date suggests that the ordinal technique has approximately the same reliability as the ordered metric technique; a comparison between the reliabilities of the interval and ordered metric techniques hardly seems meaningful in view of their different measures of internal consistency (see immediately below).

(c) Reliability

One of the important features of Shapiro's original technique and all generalized personal questionnaire techniques is that, unlike the majority of conventional questionnaires, they have a built-in test of the internal consistency of the patient's responses. For the ordinal and ordered metric techniques this test is provided by Slater's statistic i, whereas for the interval technique it is given by the final analysis of variance.

A specification of the reliability of a particular questionnaire item constructed for a particular patient can be obtained from this test of internal consistency. For the ordinal and ordered metric techniques this specification is given simply by the proportion of the patient's responses which are consistent (i.e. the difference between unity and the proportion, given by summing i over all occasions and dividing by the total number of responses, of inconsistent responses). For example, for the three items in Table 14.12, the internal consistencies are

(a)   Stammering $1 - \dfrac{2}{5 \times 32} = 0.988$

(b)   Sticking up for self $1 - \dfrac{9}{5 \times 26} = 0.931$

(c)   Tenseness $1 - \dfrac{1}{5 \times 18} = 0.989$

For the interval technique, this specification is given by the standard error of estimate of the occasion values, obtained from the final analysis of variance by taking the square root of the residual variance estimate divided by the number of statements. For example, for the two items in Table 14.17 and 14.18, these standard errors are

(a)   Having a rub with a boy $\sqrt{\dfrac{185.9968}{5}} = 6.099$

(b)   Having a go with my wife $\sqrt{\dfrac{133.254}{5}} = 5.162$

Lack of independence   There is, however, a special problem which arises here. It is that sets of responses from the same patient cannot be guaranteed to be independent (as required by analysis of variance) in the way that, say, sets of observations each from a single subject can. It is first of all worthwhile stating the problem precisely, since it is frequently misunderstood in a way which exaggerates it. The technique of analysis of variance does not require that all the observations shall be entirely unrelated. Rather, it assumes that each observation is the sum of a number of components, all but one of which represent the effects of the treatments being applied, and may therefore be related, while the remaining one represents random errors: it is these error components which are assumed to be independent of one another. In psychological terms, what this means in the present context is that the fact that all the responses are made by a single patient, in sets on different occasions, does not by itself constitute any violation of the assumption of independence, since if, for example, the intensity of the symptom decreased from one occasion to the next by a certain amount, then all the patient's difference judgements should decrease by just that amount: thus the judgements made by any patient responding relatively consistently may be expected to be correlated. The independence assumption would only be violated if the very fact that the patient had responded in a certain way to one statement on a particular occasion by itself influenced his response to another statement on that occasion, or to any statement on a subsequent occasion. In brief, the assumption of independence is not violated if the responses are influenced by the statements or the current level of the symptom, and is only violated if the responses influence one another.

Thus, although this problem is an important one, its importance should not be exaggerated; violations of independence will not occur necessarily or automatically with sets of responses from a single subject, but only if the responses actually contaminate one another. There are three points which must be made. Firstly, although the problem has been raised in the context of interval questionnaires, it is not peculiar to either analysis of variance or personal questionnaires: it arises equally with any experiment whatsoever involving repeated observations on the same subject, analysed by almost any statistical technique whatsoever, for virtually all statistical tests make some assumption of independence. Secondly, the problem is extremely difficult, and so far unsolved: some work on it in the context of analysis of variance has been carried out (for a review, see Scheffé, 1959), but no definitive solution has yet been attained. Thirdly, therefore, there seem to be only three possible courses of action open to the psychologist:

1. He may abandon all attempts to carry out experiments with a single subject until a solution is available.
2. He may carry out such experiments, but strictly abjure any use of statistical tests. However, this is hardly practicable, since unless his results are much more clear-cut than experience gives us any right to expect, they will always be open to the objection that they arose by pure chance, an objection which can only be met by the use of statistical tests.
3. He may continue to carry out single-subject experiments, employing conventional statistical tests, in the hope that the effects of any possible violations of the independence assumption will not be too serious.

The present writer prefers the third alternative.

(d) Validity

Generalized personal questionnaire techniques are concerned with the complaints which a patient utters. There are thus two things which they might be thought of as attempting to scale: firstly, the patient's actual complaints or, secondly, the things of which he complains. The distinction is crucial, for the latter are subjective states of mind which are not, in the present state of out knowledge, directly accessible, objectively definable or scalable, so that the relationship between their intensity and the patient's complaints is not known. Thus in this case the question of validity is completely unanswerable.

Hence it is better to think of these techniques as attempting to scale the patient's actual complaints, as communicational devices which try to enable him to quantify the statements he makes, to the psychiatrist or psychologist, about his current state of mind in a way which the relative imprecision of everyday language does not permit. In this case the question of validity does not arise; the only question is whether the patient can use the communicational device effectively and coherently, and this is answered by the internal consistency tests of reliability just discussed. If there is a high degree of internal consistency, then successful communication has been achieved; if not, then it has not.

To be sure, personal questionnaires, as communicational devices, can be used, either deliberately or as a further feature of the patient's illness, to transmit false information just as well as true: but the psychiatrist or psychologist will not necessarily accept the information transmitted by them at face value, any more than he necessarily accepts at face value any other communications from the patient. For example, in the writer's clinical judgement, the information trans-mitted by the patients of *Examples 1* and *4* was true, while that of the patient of *Example 5* was accepted with a great deal of reserve. (However, even false information can be clinically meaningful and valuable.) The relationship between patients' personal questionnaire responses and a psychiatrist's judgements has been studied by Shapiro and Post (1974).

(e) Applications of generalized personal questionnaire techniques

With a few exceptions, all applications of personal questionnaire techniques have employed Shapiro's three-statement ordered metric item. The exceptions are the work of Mulhall (1976) who, as mentioned above (Section 14.7c), has devised an easy-to-use method of constructing six- and eight-statement ordered metric questionnaires and given examples of its application in clinical practice, and of a small number of other workers who have constructed standard (i.e. non-personal) questionnaires using ordered metric items. Ingham (1965) used a questionnaire consisting of modified seven-statement items to scale backache, fatigue, anxiety, headache and depression, and modified five-statement items to scale sympathy for these symptoms, in a survey of symptoms and attitudes in a sample from the population of a rural area in South Wales. Sambrooks and MacCulloch (1970, 1973) and MacCulloch and Sambrooks (1974) employed the Sexual Orientation Method (S.O.M.), a questionnaire of hetero- versus homosexual orientation, consisting of ordered metric items (modified from the scale devised by Feldman and others, 1966), as an adjunct to investigation of the behaviour therapy of homosexuals. Harbison and others (1974) have constructed an ordered metric questionnaire of sexual interest (SIN) as part of an ongoing research project into the treatment of sexual disorders.

However, although part of the personal questionnaire methodology does indeed provide a way of constructing standard questionnaires which may thereby have the advantage of a better rationale than other types, this does not exploit the full potential of the technique, which stems from the fact that it enables the user to construct a special questionnaire for each individual patient. Here it is to be regretted that researchers have scarcely begun to take advantage of the possibilities of the method: over the fifteen years since it was first published there has been an average of less than two studies per year applying it, the (one) name of two authors accounting for about half the literature. This failure may be due in part to insufficient appreciation of the importance of the research strategy developed by Shapiro (Shapiro and Ravenette, 1959; Shapiro, 1961c, 1963, 1964a, 1964b, 1966, 1972, 1975) which led to the invention of personal questionnaires. It is the more disappointing because such research as has employed the technique has more than amply justified its use.

The literature may be divided into those studies which deal with the effects of a single treatment upon reports of symptoms and those which compare the effects of different treatments. In most of the former the treatment has been psychotherapy of one kind or another (and psychotherapy has been one of the treatments compared in most of the latter), although there have been a few exceptions. Thus McPherson and LeGassicke (1965), in a study which should be a model for drug research, compared the effects of Wy 3498 (a sedative analogous to chlordiazepoxide) with those of a placebo on a single patient suffering from a variety of anxiety and obsessional symptoms. Again, Nias and Shapiro (1974), in an investigation of the effects of polarization in two depressed patients, found that the symptoms of one tended consistently to improve after one hour of small scalp negative current, whereas those of the other tended to improve with positive current. Turning now to the psychotherapy studies, Walton and McPherson (1968), in an investigation of various aspects of group-analytic therapy, found a complicated pattern of symptom fluctuation in individual patients, which contrasted strikingly with clear and regular changes in group processes; this complexity of the effects of psychotherapy is a recurring theme in personal questionnaire studies. Mitchell (1969a, 1969b) investigated the long-term, immediate and delayed effects of directive counselling on a single subject (both papers report the same experiment) and whether the earlier counselling sessions were followed by more improvement than later ones. His finding of delayed and overall improvement despite no immediate effects is another recurring theme in personal questionnaire studies of psychotherapy, as opposed to other types of treatment such as desensitization, relaxation and occupational therapy. Thus Hobson and Shapiro (1970) administered personal questionnaires to seven patients in a psychotherapy community ward immediately before and after individual psychotherapy sessions, and immediately before and after a one-hour comparison period, and found more changes 'for better or worse' (more for worse) during psychotherapy and no evidence of a trend towards improvement over a twelve-week period which was intensively studied, although there was some suggestion of improvement after a further 5½ months of treatment. These findings were confirmed in a single case study by Shapiro and Hobson (1972), who also reported insignificant correlations between personal questionnaire changes and ratings of therapist Accurate Empathy, Non-possessive Warmth, Genuineness and the patient's Depth of Self-Exploration (Truax and Carkhuff, 1967); the greatest amount of variation in symptoms was for tension and depression, confirming a personal communication by Shapiro. Further confirmation of these findings was obtained in a study of seven members of an interpretative psychotherapy group by Shapiro and others (1975), who also found evidence of an inverse relationship between changes during sessions and changes over the period of the study; the authors were able to point out clinically meaningful relationships between the personal questionnaire data and verbal behaviour and repertory grid measures of the patients.

The topic of the relationship between personal questionnaire data and other scales makes it appropriate to mention here a small group of studies which have considered systematically a problem usually overlooked or less thoroughly treated, namely the relationship between symptoms *within* a personal questionnaire. Slater (1970, 1974) applied principal components analysis to the data of the first of the three patients of Shapiro (1969, see below) and extracted two components, the first appearing to scale aggravation or relief of symptoms in general, the second being described provisionally as one of internal/external reference (see Shapiro, Litman and Hendry, 1973, below): there was a strong parabolic relationship between the two components. Pritchard, Rump and Grivell (1972) preferred a modified version of McQuitty's (1961) cluster analysis in analysing the data of a manic-depressive patient treated by E.C.T. and a variety of drugs, as did Rump (1974), who reanalysed Slater's (1970) example, obtaining four symptom clusters. The relative merits of the two methods of analysis are discussed by Rump (1974) and Slater (1974).

Turning now to studies comparing the effects of different treatments, there is one preliminary remark which it ought not to be necessary to make, namely that experiments of this type should be designed in such a way that the effects of the different treatments can be separated out from the data. Thus little can meaningfully be concluded from the results of Fox and Di Scipio (1968), who attempted to eliminate 'phobic' responses to heterosexual activity and by positive conditioning to reestablish heterosexual responses in a married homosexual patient, but overlapped these two treatments, or from those of Pritchard, Rump and Grivell (1972), where there was multiple confounding and overlapping of E.C.T. and various drug treatments. An appropriate design was indicated by Shapiro, Neufeld and Post (1962), who applied, in a balanced order, systematic desensitization to the phobias and rational psychotherapy mainly to the depressive symptoms of a single patient: in this preliminary report the authors noted that 'of the 13 symptoms or symptom groups which disappeared, remission of 11 had been preceded by a specific series of treatments, suggesting a considerable degree of functional autonomy'. Shapiro, Marks and Fox (1963) used a more elaborate, five-variable, factorial, design, the factors being questionnaire items, non-directive psychotherapy versus systematic desensitization, immediate versus delayed measurements, first versus second halves of the experiment and midweek versus weekend: their main findings were that '(i) each form of treatment produced immediate beneficial effects; (ii) the delayed effect of the 'rational—training' sessions appeared to be that of relapse, and that of the 'non-directive' to be variable though with an average effect of improvement; (iii) previous findings of the relatively independent fluctuations of phobic and depressive symptoms were confirmed'. In a study of three further depressed patients, Shapiro (1969) administered personal questionnaires immediately before and after treatment by psychotherapy and occupational therapy: for the first two patients, the large majority of their symptoms showed immediate treatment improvements, there being no clear difference between the immediate effects of the two kinds of treatment sessions. (Results from the third were unclear, since she frequently refused treatment sessions.) Inconsistency from patient to patient of the effects of psychotherapy emerged in two further patients treated by behaviour modification and non-directive psychotherapy (Shapiro, 1972) where, in contrast to previous results,the feeling of depression showed a higher frequency of within-session improvements during the behaviour modification sessions than during the psychotherapy interviews. The importance of considering different symptoms separately was highlighted in a study of five patients with longstanding phobias and depression by Yaffe (1972), who found that 'the depressive symptoms showed a consistent reduction in the reported intensity during behaviour therapy, whereas no such consistent effect was observed among these symptoms during [non-directive] psychotherapy. The phobic symptoms showed a consistent reduction . . . during both treatments'. The data of the second of the two patients of Shapiro (1972) were further analysed by Shapiro, Litman, Nias and Hendry (1973), and those of the first, together with another patient, by Shapiro, Litman and Hendry (1973), who distinguished between two types of symptom statement, 'referential', where the feelings were related to circumstances outside the subject himself and to times other than the present, and 'non-referential', where they were not related to anything but the subject in the immediate present. Referential states produced less frequent reductions of intensity than non-referential states; the results for increases of intensity were similar for one patient and inconsistent for the other. Finally, in this group of studies, some evidence for the possibility, raised by Shapiro and Hobson (1972), that different types of psychotherapy, interpretative and supportive, might have different effects, was found by Shapiro and Shapiro (1974), but the difference was confounded by a number of other factors.

Two conclusions, adumbrated earlier in this section, appear to be justified by the literature surveyed. Firstly, personal questionnaires have demonstrated their value in suggesting lines of research and acquiring important new knowledge. Secondly, the surface of the field of research accessible in this was has, as yet, barely been scratched.

(f) Other types of personal questionnaire technique

It would be wrong to conclude without pointing out that the types of technique described here are not the only possible ones. Others, corresponding to other types of scale, are at least theoretically feasible. For example, a ratio type is conceivable, very similar to the interval type, but in which the patient is asked to make ratio judgements (whose *logarithms* are analysed). It would, however, suffer from the limitation of being only applicable with unipolar dimensions, for if a symptom could assume values on both sides of a natural, non-arbitrary zero point (such as 'Don't know'), then negative ratios could arise which would prevent use of logarithms. Again, a technique could be devised corresponding to a type of scale described by Luce (1956) and called a semi-order, which specifies, for pairs of objects, which is *discriminably* greater than the other, or whether they are not discriminably different. However, those types of technique are the only ones which have been tried out in practice.

Notes

1. The word 'measurement' is also frequently used in the literature as synonymous with 'scaling'. However, this gave rise to a controversy between psychologists and other scientists, and certainly there do seem to be differences in the meaning of the former term in psychology and, say, physics (Ellis, 1966). Moreover, 'measurement' is sometimes used by psychologists in a rather grandiose way, e.g. 'measurement of personality'. For these reasons the more modest term 'scaling' will be preferred here.
2. It is special in that it used just three reference statements, and also in certain other respects mentioned below (Section 14.2c).
3. For an account of ordinal, interval and ratio scales, see Stevens (1946, 1951).
4. For an account of partial ordinal and partial ordered metric scales, see Coombs (1951).
5. For an account of ordered metric scaling, see Coombs (1950, 1951).
6. For further data on this patient see Phillips (1964).

References

Ament, W. (1900). 'Ueber das Verhältniss der ebenmerklichen zu den übermerklichen Unterschieden bei Licht- und Schallintensitäten'. *Phil. Stud.*, **16**, 135–196.

Cherry, C., and Sayers, B. McA. (1956). 'Experiments upon the total inhibition of stammering by external controls, and some clinical results'. *J. psychosom. Res.*, **1**, 233–246.

Coombs, C. H. (1950). 'Psychological scaling without a unit of measurement'. *Psychol. Rev.*, **57**, 145—158.

Coombs, C. H. (1951). 'Mathematical models in psychological scaling'. *J. Amer. statist. Assoc.*, **46**, 480—489.

Eichenlaub, J. E. (1964). *The Marriage Art*, Mayflower—Dell, London.

Ellis, B. (1966). *Basic Concepts of Measurement*, Cambridge University Press.

Feldman, M. P., MacCulloch, M. J., Mellor, V. and Pinschof, J. M. (1966). 'The application of anticipatory avoidance learning to the treatment of homosexuality. III. The Sexual Orientation Method'. *Behav. Res. Ther.*, **4**, 289—299.

Fox, B., and Di Scipio, W. J. (1968). 'An exploratory study in the treatment of homosexuality by combining principles from psychoanalytical theory and conditioning: theoretical and methodological considerations'. *Brit. J. med. Psychol.*, **41**, 273—282.

Guilford, J. P. (1954). *Psychometric Methods*, 2nd ed. McGraw-Hill, New York.

Harbison, J. J. M., Graham, P. J., Quinn, J. T., McAllister, H. and Woodward, R. (1974). 'A questionnaire measure of sexual interest'. *Arch. sex. Behav.*, **3**, 357—366.

Hobson, R. F., and Shapiro, D. A. (1970). 'The personal questionnaire as a method of assessing change during psychotherapy'. *Brit. J. Psychiat.*, **117**, 623—626.

Ingham, J. (1965). 'A method for observing symptoms and attitudes'. *Brit. J. soc. clin. Psychol.*, **4**, 131—140.

Kelly, G. A. (1955). *The Psychology of Personal Constructs*, W. W. Norton & Co., New York.

Kendall, M. G. (1948). *Rank Correlation Methods*, Griffin, London.

Kendall, M. G., and Smith, B. B. (1939). 'On the method of paired comparisons'. *Biometrika*, **31**, 324—345.

Luce, R. D. (1956). 'Semiorders and a theory of utility discrimination'. *Econometrica*, **24**, 178—191.

MacCulloch, M. J., and Sambrooks, J. E. (1974). 'Sexual interest latencies in aversion therapy: a preliminary report'. *Arch. sex. Behav.*, **3**, 289—299.

McPherson, F. M., and LeGassicke, J. (1965). 'A single-patient self-controlled trial of Wy 3498'. *Brit. J. Psychiat.*, **111**, 149—154.

McQuitty, L. L. (1961). 'Elementary factor analysis'. *Psychol. Rep.*, **9**, 71—78.

Miller, G. A. (1947). 'Sensitivity to changes in the intensity of white noise and its relation to masking and loudness'. *J. acoust. Soc. Amer.*, **19**, 609—619.

Mitchell, K. R. (1969a). 'Shapiro's single case repeated-measure design applied to the individual client in counselling'. *Australian Psychologist*, **4**, 20—36.

Mitchell, K. R. (1969b). 'Repeated measures and the evaluation of change in the individual client during counseling.' *J. counsel. Psychol.*, **16**, 522—527.

Mulhall, D. J. (1976). 'Systematic self-assessment by P.Q.R.S.T.'. *Psychol. Med.*, (in press).

Newman, E. B. (1933). 'The validity of the just noticeable difference as a unit of psychological magnitude'. *Trans. Kansas. Acad. Sci.*, **36**, 172—175.

Nias, D. K. B., and Shapiro, M. B. (1974). 'The effects of small electrical currents upon depressive symptoms'. *Brit. J. Psychiat.*, **125**, 414—415.

Osgood, C. E., Suci, G. J., and Tannenbaum, P. H. (1957). *The Measurement of Meaning*, University of Illinois Press, Urbana.

Phillips, J. P. N. (1963). 'Scaling and personal questionnaires'. *Nature*, **200**, 1347—1348.

Phillips, J. P. N. (1964). 'Techniques for scaling the symptoms of an individual psychiatric patient'. *J. psychosom. Res.*, **8**, 255—271.

Phillips, J. P. N. (1966). 'On a certain type of partial higher-ordered metric scaling'. *Brit. J. math. statist. Psychol.*, **19**, 77—86.

Phillips, J. P. N. (1967). 'A procedure for determining Slater's *i* and all nearest adjoining orders'. *Brit. J. math. statist. Psychol.*, **20**, 217—225.

Phillips, J. P. N. (1968a). 'A note on the scoring of the Sexual Orientation Method'. *Behav. Res. Ther.*, **6**, 121—123.

Phillips, J. P. N. (1968b). 'A scheme for computer scoring a Shapiro personal questionnaire'. *Brit. J. soc. clin. Psychol.*, **7**, 309—310.

Phillips, J. P. N. (1968c). *Psychological Scaling and Generalised Personal Questionnaire Techniques*, Unpublished Ph. D. thesis, University of London.

Phillips, J. P. N. (1969). 'A further procedure for determining Slater's *i* and all nearest adjoining orders'. *Brit. J. math. statist. Psychol.*, **22**, 97—101.

Phillips, J. P. N. (1970a). 'A new type of personal questionnaire technique'. *Brit. J. soc. clin. Psychol.*, **9**, 241—256.

Phillips, J. P. N. (1970b). 'A further type of personal questionnaire technique'. *Brit. J. soc. clin. Psychol.*, **9**, 338—346.

Phillips, J. P. N. (1971a). 'The investigation of irrelevant response set in ordinal and ordered metric personal questionnaries'. *Bull. Brit. Psychol. Soc.*, **24**, (No. 82), 62 (Abstract).

Phillips, J. P. N. (1971b). *A Simple Method of Obtaining a Ratio Scaling of a Hierarchy*, Paper presented to the Third Conference of the Behavioural Engineering Association, Wexford, Ireland.

Phillips, J. P. N., Hutchings, R. L., and Thompson, J. W. (1967). 'Analysis and design of certain scaling experiments'. *Nature*, **215**, 897—898.

Phillips, J. P. N., and Kenyon, P. M. (1975). 'An approach to the experimental investigation of some parameters of systematic desensitization of the individual patient'. In J. C. Brengelmann (Ed.), *Progress in Behaviour Therapy*, Springer—Verlag, Berlin.

Pritchard, D. W., Rump, E. E., and Grivell, P. D. (1972). 'Time series for symptom-clusters in a case of manic-depressive illness'. *Aust. N. Z. J. Psychiatry*, **6**, 231—237.

Remage, R. Jr., and Thompson, W. A. (1966). 'Maximum likelihood paired comparison rankings. *Biometrika*, **53**, 143—149.

Rump, E. E. (1974). 'Cluster analysis of personal questionnaires, compared with principal component analysis'. *Brit. J. soc. clin. Psychol.*, **13**, 283—292.

Sambrooks, J. E., and MacCulloch, M. J. (1970). *The Role of Response Latencies in the Detection of Faking on a Personal Questionnaire*, Paper presented at the second Annual Conference of Behavioural Engineers at Kilkenny.

Sambrooks, J. E., and MacCulloch, M. J. (1973). 'A modification of the Sexual Orientation Method and an automated technique for presentation and scoring'. *Brit. J. soc. clin. Psychol.*, **12**, 163—174.

Scheffé, H. (1959). *The Analysis of Variance*, John Wiley and Sons, New York.

Shapiro, D. A., and Hobson, R. F. (1972). 'Change in psychotherapy: a single case study'. *Psychol. Med.*, **2**, 312—317.

Shapiro, D. A., Caplan, H. L., Rohde, P. D., and Watson, J. P. (1975). 'Personal questionnaire changes and their correlates in a psychotherapeutic group'. *Brit. J. med. Psychol.*, **48**, 207—215.

Shapiro, M. B. (1961a). 'A method of measuring changes specific to the individual psychiatric patient'. *Brit. J. med. Psychol.*, **34**, 151—155.

Shapiro, M. B. (1961b). *Abbreviated Manual. THE PERSONAL QUESTIONNAIRE*, Unpublished.

Shapiro, M. B. (1961c). 'The single case in fundamental clinical research'. *Brit. J. med. Psychol.*, **34**, 255—262.

Shapiro, M. B. (1963). 'A clinical approach to fundamental research with special reference to the study of the single patient'. In P. Sainsbury and N. Kreitman (Eds.), *Methods of Psychiatric Research*, Oxford University Press.

Shapiro, M. B. (1964a). 'The measurement of clinically relevant variables'. *J. psychosom. Res.*, 8, 245–254.

Shapiro, M. B. (1964b). 'The single case in psychological research: a reply'. *J. psychosom. Res.*, 8, 283–291.

Shapiro, M. B. (1966). 'The single case in clinical-psychological research'. *J. gen. Psychol.*, 74, 3–23.

Shapiro, M. B. (1969). 'Short-term improvements in the symptoms of affective disorder'. *Brit. J. soc. clin. Psychol.*, 8, 187–188.

Shapiro, M. B. (1972). 'A direct approach to the investigation of affective disorder'. In D. Eaves (Ed.), *Recent Developments in Psychiatry*, The Royal College of General Practitioners, London.

Shapiro, M. B. (1975). 'The requirements and implications of a systematic science of psychopathology'. *Bull. Brit. psychol. Soc.*, 28, 149–155.

Shapiro, M. B., Litman, G. K., and Hendry, E. R. (1973). 'The effects of context upon the frequency of short-term changes in affective states'. *Brit. J. soc. clin. Psychol.*, 12, 295–302.

Shapiro, M. B., Litman, G. K., Nias, D. K. B., and Hendry, E. R. (1973). 'A clinician's approach to experimental research'. *J. clin. Psychol.*, 29, 165–169.

Shapiro, M. B., Marks, I. M., and Fox, B. (1963). 'A therapeutic experiment on phobic and affective symptoms in an individual psychiatric patient'. *Brit. J. soc. clin. Psychol.*, 2, 81–93.

Shapiro, M. B. Neufeld, I. L., and Post, F. (1962). 'Note: experimental study of depressive illness'. *Psychol. Rep.*, 10, 590.

Shapiro, M. B., and Post, F. (1974). 'Comparison of self-ratings of psychiatric patients with ratings made by a psychiatrist'. *Brit. J. Psychiat.*, 125, 36–41.

Shapiro, M. B., and Ravenette, A. T. (1959). 'A preliminary experiment on paranoid delusions'. *J. ment. Sci.*, 105, 295–312.

Shapiro, M. B., and Shapiro, D. A. (1974). 'Experiments on the feeling of depression'. *Brit. J. soc. clin. Psychol.*, 13, 191–199.

Singer, W. B., and Young, P. T. (1941). 'Studies in affective reaction: I. A new affective rating scale'. *J. gen. Psychol.*, 24, 281–301.

Slater, P. (1960a). 'The analysis of personal preferences'. *Brit. J. statist. Psychol.*, 13, 119–135.

Slater, P. (1960b). *The Reliability of Some Methods of Multiple Comparison in Psychological Experiments*, Unpublished Ph.D. thesis, University of London.

Slater, P. (1961). 'Inconsistencies in a schedule of paired comparisons'. *Biometrika*, 48, 303–312.

Slater, P. (1965). 'The test–retest reliability of some methods of multiple comparison'. *Brit. J. math. statist. Psychol.*, 18, 227–242.

Slater, P. (1970). 'Personal questionnaire data treated as forming a repertory grid'. *Brit. J. soc. clin. Psychol.*, 9, 357–370.

Slater, P. (1974). 'Cluster analysis versus principal component analysis: a reply to E. E. Rump'. *Brit. J. soc. clin. Psychol.*, 13, 427–430.

Stephenson, W. (1953). *The Study of Behavior. Q-technique and Its Methodology*, University of Chicago Press.

Stevens, S. S. (1946). 'On the theory of scales of measurement'. *Science,* 103, 677–680.

Stevens, S. S. (1951). 'Mathematics, measurement and psychophysics'. In S. S. Stevens (Ed.) *Handbook of Experimental Psychology*, John Wiley and Sons, New York.

Stevens, S. S. (1959). 'Sic transit gloria varietatis?'. *Contemp. Psychol.*, 4, 388–389.

Stevens, S. S. (1961). 'Towards a resolution of the Fechner-Thurstone legacy'. *Psychometrika*, 26, 35–47.

Torgerson, W. S. (1958). *Theory and Methods of Scaling*, John Wiley and Sons, New York.

Truax, C. B., and Carkhuff, R. R. (1967). *Toward Effective Counseling and Psychotherapy*, Aldine, Chicago.

Walton, H. J., and McPherson, F. M. (1968). 'Phenomena in a closed psychotherapeutic group'. *Brit. J. med. Psychol.*, 41, 61—72.

Wolpe, J. (1958). *Psychotherapy by Reciprocal Inhibition*, Stanford University Press.

Woodward, R., McAllister, H., Harbison, J. J. M., Quinn, J. T., and Graham, P. J. (1973). 'A comparison of two scoring systems for the S.O.M.'. *Brit. J. soc. clin. Psychol.*, 12, 411—414.

Yaffe, O. (1972). *Effects of Behaviour Therapy and Psychotherapy upon the Co-variation of Phobic and Depression Symptoms in Individual Psychiatric Patients*, Unpublished Ph. D. thesis, University of London.

PART IV

EPILOGUE

15

ON THE PLACE OF GRID TECHNIQUE IN
PSYCHOLOGY

15.1 Systems for explaining psychological phenomena

Psychology covers an enormous range of phenomena: the characteristics and behaviour of human beings of all kinds individually and in general, their relations with one another and with their surroundings, and also the whole of the animal Kingdom similarly. Within this vast domain the research worker is often satisfied to single out a small patch and confine himself to cultivating it. The teacher, on the other hand, is liable to be placed under an obligation to encapsulate the whole subject into a one-year subsidiary course (or an even shorter one) for students whose main interests are in other matters. To him any system of explanation that purports to be universal is likely to be welcome.

Many such systems have appeared in the history of psychology, beginning with associationism. Psychoanalysis, trait psychology, behaviourism and personal construct theory are more recent ones. These systems offer alternative interpretations of the phenomena, any of which may be employed heuristically for the solution of a particular problem. Empirical eclecticism advocates this use of them, but it is not adopted by everyone; each system is liable to be supposed by some people to be objectively true.

Most theoretical systems have some axiomatic basis and most of the ones considered here incorporate axioms which are variants of the principle of causality, namely that every event has a sufficient cause. Psychological determinism is postulated in one form by psychoanalysis, which seeks out psychological origins for psychological disturbances, and in another form by trait psychology, which attributes differences in peoples' conduct to differences in their personalities. The stimulus—response formula of the behaviourists is unmistakeably deterministic too.

Kelly's system, in contrast, is teleological. He formulated an axiomatic basis for it in his fundamental postulate,

A person's processes are psychologically channelised by the ways in which he anticipates events,

and a set of corrollaries, of which the first is

A person anticipates events by construing their replications.

Notes are added to explain the terms used. The whole exposition expresses the conviction that people act spontaneously and direct their thoughts and actions towards achieving their purposes.

A discussion of the distinction between the two modes of explanation seems unavoidable.

15.2 Determinism and teleology

Determinism has not been as successful in application to psychology as to some other branches of science, in spite of very ingenious efforts to apply it rigorously. Terms with teleological implications are not easily excluded from deterministic systems altogether. Common speech contains a virtually inexhaustible supply, and some, like 'need', 'use', 'in order to' or simply 'to' and 'for', slip so easily into any kind of discourse that their implications are not examined critically. Others, like 'instinct', 'drive', 'motivation' and 'conation', are officially admitted, for a while at least, in systems that profess to be deterministic. They propose a causal explanation for processes which appear to be directed towards goals by postulating forces which, when released, cause events to proceed in predetermined directions. There are similar teleological implications in other terms, such as 'frustration', 'inhibition' and 'sublimation', which imply interrupting processes directed towards goals.

It is easy to understand why teleological explanations are considered objectionable. How can the possible consequences of an event be included among its actual causes? Only other events that preceded it can have caused it and it, in turn, is the cause of the consequences that follow from it. A favourite example of physical causation is the movement of balls on a billiard table: the speed and direction imparted to one is determined by the impact of another on it. Each event in the series of repercussions is the effect of the preceding ones and occurs in accordance with laws that can be formulated and applied accurately. Psychologists who take physics as their model hope that eventually they may be able to explain the course of a series of psychological events with the same accuracy.

The account of causation given originally by Aristotle is more comprehensive: it includes prime causes and final causes. In the case of the billiard balls the prime cause is the stroke from the player's cue which sets the first ball in motion and the final cause is the position they occupy when they have stopped moving. If the player is sufficiently skilful this will be precisely the one he intended for them originally. A similar account can be applied to manufacturing processes, among others. The final cause is the end product; capital and labour are invested in the process in anticipation that there will be sufficient effective demand for the product when it comes onto the market to justify the investment.

Aristotle's account supports an anthropomorphic interpretation of the history of the universe. Its prime cause is believed to be the act of a Divine being and its final cause the accomplishment of His purpose in the beginning. This interpretation is in accordance with the religious beliefs of most of humanity throughout almost all recorded time and is linked with the moral doctrine that people should try to act in accordance with the Divine will.

It became discredited among the rationalists of the eighteenth century. Hume attacked the causal principle itself. Indeed, it is obvious that the proposition that every event has a sufficient cause cannot be proved empirically. Kant argued in reply that space, time and causality are necessary formal components of experience. Space and time are a mental reference system enabling objects to be recognized as the sources of sensations, and causality enables the connections between events to be understood.

Starting at whatever points we like in time we can conceive of other times before and after; or setting whatever bounds we like to a region in space we can conceive of greater expanses beyond it and smaller regions within it. The human mind can accept explanations in terms of proximate causes, but not in terms of prime or final causes. It demands to know what came before and what will come after. It reaches out enquiringly in all directions and knows no limit. Nineteenth century science extended the known frontiers of space and time prodigiously and in the present century the process is still continuing.

15.3 Freedom of choice

Grid technique can be employed without accepting personal construct theory in its entirety. There is no need to postulate that psychological processes are always concerned with anticipating events or that construing is the only operation they employ. A sufficient basis for it is provided by the two propositions.

(a) Circumstances frequently permit some freedom of choice.
(b) Construing the alternatives can help in choosing between them.

Logically speaking both are particular propositions, not universal ones, and they can easily be verified from daily experience.

The problem of reconciling determinism with teleology cannot be avoided in discussing the place of grid technique in psychology merely by conceding that its two basic propositions are not universal. If the principle of causation is universal there can be no circumstances that permit freedom of choice; conversely, asserting that there are any circumstances whatever under which freedom of choice is possible denies the universality of the principle. To reconcile the teleological and the determinstic modes of explanation we must satisfy ourselves that the principle has some limitations as well as that there are some circumstances that permit freedom of choice. Let us consider an example.

A man leaves home to keep an appointment some distance away. He may go there by train; several coach services are also available to take him there; or if he goes by car there are several routes to choose from. He decides to go by car along one of these routes. Turning a bend as he goes downhill he runs into a patch of oil on the road, goes into an uncontrollable skid and ends up in hospital.

In a strictly deterministic account of the man's behaviour at the time of the accident is might be sufficient to begin by recording that he was driving his car at a certain speed in a certain direction when he reached the patch of oil. But if we ask why he was doing so we must either accept a teleological explanation or none — he was travelling that way because he was intending to do something else later, namely keep an appointment. If we enquire further why he came that way and not another, we trace the sequence of events back to the point in time when he took the decision. At that moment he was free to choose. The situation was indeterminate, suspended without any outcome until he decided what to do. With that it became fully determined.

A time like this, when alternative courses of action are possible and a decision is needed, is a critical moment. It may not be perceived as such: the subject may have been in similar situations before and followed a habitual course of action without reviewing the other possibilities Or social conventions may dictate a particular course. If he is aware of the need for a decision he may make it on the spur of the moment in an arbitrary way intellectually on a par with tossing a coin, or he may evade the responsibility for taking it by acting on someone else's advice, or if he has visualized the possibility of being in such a situation he may have considered the alternatives previously and had a plan already worked out. Then when the moment comes the plan is put into effect automatically.

Even if all the alternatives open at a critical moment may not always be deliberately evaluated, some attempt at evaluation is often made, and then the criteria for choice will usually include considerations of future advantages. The subject, at least momentarily a free agent, gazes into his crystal ball, his private universe, to see what the future will hold if he does this or that. Time travel is possible for him within it, though it may be impossible for anyone in reality. At the critical moment microcosm and macrocosm interlock and teleological considerations can affect the course of events.

The question 'Why?' changes in meaning when transferred from the macrocosm to the microcosm. When applied to the macrocosm it enquires into causes; when applied to the microcosm we may say that it enquires into reasons, using the word broadly to include reasons that are not entirely reasonable.

It may be argued, on the contrary, that even if events sometimes appear to occur in this way there is no need to suppose that the mental processes of which a subject is aware in making a decision actually affect the decision he makes. Being the person that he is, the decision he makes is the decision he is bound to make, and the mental processes of which he is aware are mere epiphenomena, channels that light up in the circuit from stimulus to response.

This theory is best regarded as one that offers a convenient method of approximating to the facts. Many difficulties would be encountered in trying to fit it to them exactly, and the methods in use for assessing it indirectly — tests, questionnaires, interviews, etc. — involve confronting him with situations which leave him free to respond in different ways and noting the choices he makes. It would presumably be difficult for him to express his personality in a situation that allowed him no freedom.

If procedures which allow him some freedom are used to assess his personality the theory is bypassed and replaced by the assumption that people act consistently on the whole, so that evidence on what affects their decisions on particular occasions is unnecessary. Noting what choices the subject makes in situations that are controlled experimentally and inferring what choices he is likely to make at a critical moment may often have advantages over waiting till the moment comes to find out his reasons for acting as he does. The assumption may be considered acceptable if approximations based on it reach a required standard of accuracy; but freedom of choice is not disproved.

Though dramatically critical moments may not occur frequently, every moment of the day may be regarded as more or less critical. The question, 'Is there anything else I could be doing at this moment instead of what I am doing?', is one that rarely has to be answered with an unqualified 'No'. Situations are seldom, if ever, completely determined; they leave some margin, if only small, for the exercise of choice. The naturally gregarious rat, isolated and incarcerated in his Skinner box, may take a break from his predetermined schedule of bar-presses to groom himself, explore the confines of his little cell, defecate, urinate or just have a nap, thereby bedevilling the best-laid plans of the behaviourist (Yaktin and Slater, 1965). Human beings, too, may find a sceptical attitude towards necessity worth cultivating.

If no decision is taken at the critical moment external events may alter the situation, some of the options may be eliminated or the opportunity for choosing may be lost altogether. Otherwise the situation may drag on unchanged. Anyone who has held a subordinate position in a large organization or who has applied to a research committee for a grant must know how debilitating delays in reaching decisions can be.

The 'causes' for which men combine are objectives desired by many and too difficult for anyone to achieve unaided. Groups of all sizes are formed, with similar or divergent aims and policies, and it is through the collaboration and conflict between them that teleological considerations obtrude most dynamically, for better or worse, into the course of history. The aims and policies are decided among the people with whom the power rests, usually in consultations where alternatives are considered. Prodigal resources accumulated over geological time are available for men to use or misuse at choice.

Humanity must seem very much alone in its uncertain journey through space and time if it cannot count on Divine guidance and its future depends wholly on whether the decisions taken at critical moments by the people in power are wise ones. Each century, as power becomes more concentrated, the dangers of unwise decisions increase. History is unrepeatable.

The processes involved in evaluating alternatives and reaching decisions form a proper and important subject for psychology to study, and one for which grid technique is appropriate. It is not a suitable technique for predicting and controlling behaviour mechanistically, but if it can help men make comprehensive, realistic assessments of the options before them and reach their own conclusions deliberately it may perform another kind of service.

15.4 Defending a system

When a teacher commits himself to one system for explaining psychological phenomena and advocates it zealously and persuasively, his career like him, tends to become type-cast. He acquires a public image which may help him in many ways but may also hinder him in some. It is easy for him to become a member of an establishment where his views are already known and accepted as orthodox, but his freedom to change his mind may become restricted. When he has opportunities to initiate or supervise researches he may be inclined to favour ones that can be expected to produce results confirming the system he advocates. In this way unimaginative repetitive research work may be given priority and it may even happen that results which conflict with the system remain unacknowledged. Storms and Sigal (1958) have documented one instance in detail.

Moreover, statistics can be mobilized to defend an established system, presenting an explanation that only accounts for a small fraction of the phenomena as if it applied to the whole. Suppose a hypothesis deduced from the system implies that one variable depends on another; an experiment is carried out to measure the concomitant variation of the two variables in a series of cases and a significant correlation is found. The hypothesis will be regarded as confirmed and the system as validated. How large the correlation needs to be to qualify as significant depends on the length of the series. In a series of forty cases a correlation of 0.5 is highly significant $(P < 0.001)$. The hypothesis only accounts for 25 per cent. of the variance of the supposedly dependent variable, yet no concern is felt; the remaining

75 per cent. is written off as attributable to errors of observation. If the series is longer or a lower level of significance is accepted, the balance attributable to error can be larger. Taking $P < 0.05$ as significant the hypothesis only needs to account for 10 per cent. of the variance in a series of forty cases.

The research psychologist who adheres to the standard behaviour pattern, formulating a hypothesis and setting up an experiment to obtain support for it, and concludes that his results are statistically significant (presuming that they are) should feel under an obligation to add that his hypothesis fails to account for 80 or 90 per cent. of the observed variation, or whatever the residual percentage is.

It is very seldom that a psychological experiment is designed to provide an unbiased estimate of error variance. The assumption that the variation not explained by the hypothesis is negligible is generally an arbitrary one made for convenience; it excuses the experimenter from examining his data more closely. Supplementary or qualifying evidence is liable to be overlooked. At present psychologists are inclined to approach their data in much the same way as archaeologists at the beginning of the last century approached their digs, turning them over only in search of treasure and paying no attention to clay tablets, pottery or kitchen middens. Fortunately for psychologists, their losses through similar oversights are not so expensive and irretrievable.

Now that computers allow larger amounts of information to be scanned quickly and easily through to the minutest detail, the cosiness of simple universal explanatory systems may seem less enticing than it did a generation ago.

15.5 The hypothetico-deductive method

An oversimplified notion of the hypothetico-deductive method may be one of the reasons why large parts of the evidence from an experiment are often considered not worth examining. The procedure followed is to formulate a hypothesis before the experiment is carried out, and to examine the results only to find whether a significant amount of the evidence supports it. If so, the hypothesis is considered confirmed; if not, some explanation for the failure must be found.

This appears to be how hypotheses ordinarily function in perception. Goldstein (1962), following Bruner (1941) and Postman (1951), defines a perceptual hypothesis as a highly generalized state of readiness to respond selectively to certain classes of events in the environment. It results from past experience in similar circumstances; associated motives are aroused and attention turns in a particular direction. The perceptual input is received as confirming or refuting the hypothesis. The stronger the expectation the greater its likelihood of recurring in similar situations, the less the amount of appropriate information necessary to confirm it and the more the amount of inappropriate evidence necessary to discredit it. For example:

It is 2.30 a.m. Jack wakes with a start; Jill's elbow is jabbing into him.
Jack (indistinctly). What's the matter?
Jill There's a burglar downstairs. I can hear him.
Jack (arguing to save having to move). What you can hear is a sound. You suppose there's a burglar.

Jill Don't be ridiculous. Sounds don't just happen. Somebody makes them. It must be a burglar. Listen!

   (Silence reigns)

Jack I can't hear anything.

Jill He's keeping quiet now. Sssh!

   (Finally a faint noise, off)

Jill There he is! What did I tell you?

Exit Jack from bed, reluctantly. Though Jill's reasoning may be shaky, her conclusion could still be right. The proposition, 'There's a burglar in the house' is one that demands experimental, not logical, verification.

Goldstein's may be a reasonable account of how the mental activity of hypothesis-making normally functions in the panic of everyday human life, but it does not conform with the requirements of logic. To prove a hypothesis by the hypothetico-deductive method it is necessary to disprove its contrary.

Any proposition P has a contrary $\bar{P}$ (not P) which is true when P is untrue and untrue when P is true. The universe of discourse is divided into two sets, as shown in the truth table below, where the truth value 1 stands for true and 0 for untrue:

Set	P	$\bar{P}$
(a)	1	0
(b)	0	1

Set (a) is the set of instances where P is true and $\bar{P}$ untrue, and set (b) the set where P is untrue and $\bar{P}$ true. One of the two sets may be empty.

The truth value of a proposition may be ascertained in various ways: directly from observation, by syntactical deduction from another proposition or by material implication. For instance, it can be deduced syntactically that a man must have certain legal qualifications if he has been called to the bar. Or suppose someone goes to boil an egg and finds that it floats when he puts it in the water; he may conclude that it is bad and decide to get rid of it without cracking it, or he may crack it to see — arguing by material implication in the former case or by direct observation in the latter.

The argument by material implication involves two propositions: P, the egg floats, and Q, it is good, with the contraries, $\bar{P}$ and $\bar{Q}$; and a compound proposition derived from them, namely 'P implies $\bar{Q}$', which might also be stated as 'if P, then $\bar{Q}$'. The universe of discourse, eggs available for boiling, falls into four sets according to whether P and Q are true or false:

Set	Including eggs that	Truth values of P	Q
(a)	Float and are good	1	1
(b)	Float and are bad	1	0
(c)	Sink and are good	0	1
(d)	Sink and are bad	0	0

The proposition '*P* implies $\overline{Q}$' is true by material implication if set (a) is empty, i.e. if the universe contains no instances of eggs that float and are good.

Similarly Jill's argument concerns a series of nights, which can presumably be assigned to four sets: (a) nights when a sound occurs and there is a burglar in the house; (b) nights when there is a sound but no burglar; (c) nights when there is a burglar but no sound; (d) nights when there is no sound and no burglar. The proposition, 'if there is a sound there must be a burglar', is bound to be true if set (b) is empty.

The trouble with both these arguments is the same. There is no means of proving conclusively that the crucial set is completely empty. Even if there has never been a good egg yet that floated, some hen some day may perhaps contrive to lay one; even if Jill's sleep has never so far been disturbed by sounds not made by burglars, some night perhaps some other sound may waken her. The universes concerned must be considered inexhaustible. The only way to verify the conclusion is by direct observation: the egg must be cracked, the house searched. And then the conclusion reached is only a particular one, referring to the instance in question.

There is one form of argument which permits a general conclusion to be drawn from a particular instance. If syntactical deduction from some hypothesis leads to the conclusion that a set must be empty, and experimental evidence shows that it is not, proof is provided that the hypothesis is untrue. Consequently its contrary must be true. This is the logic of the null hypothesis.

The experimenter who personally believes that a proposition is true but cannot find any way of confirming it by direct observation may formulate it as a hypothesis *H* and then go on to consider what the consequences would be if it were untrue. If some proposition *P* necessarily implied by this contrary null hypothesis $\overline{H}$ can be tested experimentally and turns out to be untrue, he has found a convincing argument to confirm his hypothesis. The truth table can be drawn up as:

Set	$\overline{H}$	$\overline{P}$	$\overline{H}$
(a)	1	1	0
(b)	1	0	0
(c)	0	1	1
(d)	0	0	1

Since *P* has been proved true when $\overline{H}$ is true set (b) must be empty, and therefore any instance of $\overline{P}$ must belong to set (d). By producing an instance of $\overline{P}$ the experimenter can demonstrate that set (d) is not empty and consequently that *H* must be true. The argument reduces his task from showing that a set is empty, which cannot be done experimentally except by exhausting it, to showing that a set is not empty, which can be done by producing a single instance.

The experiment attributed to Galileo, of dropping different weights from the leaning tower of Pisa, and Lavoisier's experiment of sealing water in a container and weighing it before and after boiling it, are famous examples of such particular experiments with decisive general implications. If their logic is to be understood they must be viewed in the twilight of opinions prevalent at the time.

15.6 Mediating postulates

Psychology abounds in terms to explain phenomena by referring them to underlying entities or functions, i.e. ones that cannot be observed directly. Trait psychology nominates personal characteristics, including aptitudes and abilities, personality traits and factors of all kinds, typically intelligence; psychoanalysis employs dynamic and structural terms, such as libido, superego and id; personal construct theory is beginning to burgeon with its own set of terms, mainly structural, such as complexity and superordinacy. Conclusive experimental evidence of the existence or activity of these postulated entities is not to be had; operational definitions may point towards them but do not contain them; and they remain obscured by controversy.

Often such terms are employed to enlarge arguments of a purely speculative kind; we may then regard them as postulates with some possible explanatory power, even though their existence or activity has not been demonstrated experimentally. Some at least can also be used to mediate conclusions that can be tested experimentally, and it may then be supposed that their existence has been confirmed to some extent. Notably the postulates of trait psychology have shown their value as mediating terms in many contests.

It is convenient, for instance, to refer to a set of characteristics when considering problems in the allocation of personnel. A job analysis will be carried out to identify the ones required. Then a selection procedure will be designed incorporating methods for assessing them by tests, questionnaires, structured interviews, etc. After an experimental period the procedure will be validated by a follow-up to see how efficient it has been in matching the people to the jobs; criteria will be defined for measuring its efficiency; and formulae will be worked out for estimating a man's probability of adapting successfully to a job directly from his record in the selection procedure. (The evidence from the follow-up may also be used to revise the procedure.)

The feature of such investigations which is of interest here is that the characteristics are not concerned in the conclusions. They are bypassed. The formulae for estimating suitability for the jobs relate the information gathered in the selection procedure directly to the criteria. After they have been worked out there is no need to discuss such abstruse questions as:

Is intelligence (say) needed to perform such-and-such a job?
Is such-and-such a test really a measure of intelligence?
What do we mean by intelligence, anyhow?
Is it a unitary trait or a resultant of many traits?

Such questions are best left to the 'brains trusts'. The investigations follow the course of an argument where the characteristics are introduced as middle terms and could conceivably be replaced by others; their value as mediating postulates is demonstrated but not their existence as functional entities.

The component-space of a grid has a similar mediating value in relating the elements and the constructs together. Indeed, grid technique could be used to operate a self-selection procedure. Informants could be asked to evaluate a set of jobs and 'Would suit me best/least' could be supplied as one of the constructs. A composite diagram could then be made to compare the dispersion of the informants and the jobs in the component-space. Its principal axes would provide a convenient reference system for comparison, but they would not necessarily define any underlying psychological entities.

A postulate that mediates an argument from a set of observations to a verified conclusion must have some logical value, even if its value is uncertain. At least its value must be greater than that of a postulate that does not lead to any definite conclusion from definite evidence.

When psychological tests are used clinically a more devious line of argument may be followed, namely that certain features of the responses indicate that the mental state of the patient is disturbed in ways which are consistent with a certain type of psychiatric disorder which is likely to benefit from a certain kind of treatment. Logically it may not matter how many intermediate steps there are in an argument, 'S is an A', 'As are B', 'Bs are C', etc., that leads to the conclusion, 'S is P' but if each of the propositions is open to doubt, the longer the sequence the more doubtful the conclusion. Sometimes there may be advantages in shortening the path from observing the evidence to deciding on the treatment, avoiding the use of unnecessary postulates. This is a situation in which grid technique may help; it may even bring the presenting problem to light and indicate a promising line of treatment (see Rowe, *Explorations*, Chapter 1).

15.7 Empirical eclecticism

The student who adopts the approach of empirical eclecticism to psychology feels free to use the whole of its vocabulary. He treats different systems as different languages for expressing his ideas, thinking, for instance, of traits as constructs used by some psychologists for evaluating subjects. He is even willing to try translating teleological terms into deterministic ones, and vice versa.

When considering a problem that interests him he borrows whatever terms appear convenient for applying to it; he does so tentatively, however, because he feels free to reformulate it in other terms if they turn out to be more suitable. He is always in search of evidence to test his opinions and regards all hypotheses and postulates with philosophic doubt.

Like Newton, the father of the hypothetico-deductive method, he makes his maxim *Hypotheses non fingo* (freely translated, 'I don't go around making up hypotheses' — the choice of the derogatory repetitive *fingo* instead of the commoner *facio* cannot be overlooked). He prefers investigating directly verifiable propositions. When he finds making a hypothesis has become unavoidable he starts immediately to consider the possibility that it is untrue. He looks for any evidence he can find to test it and considers whatever techniques can be applied for the purpose. When confronted with the data from an experiment he puts aside any previously proposed hypothesis and looks at them impartially to see what they actually show. He pays careful attention to their limitations. He considers whatever other explanations can be found besides the one hypothesized. If the hypothesis finally proves entitled to some credence he allows it no more than he is sure it merits, and still feels free to consider alternatives.

References

Bruner, J. S. (1941). 'Personality dynamics and the process of perceiving'. In R. R. Blake and G. V. Ramsay (Eds.), *Perception: An Approach to Personality*, Ronald Press, New York. pp. 123—142.

Goldstein, A. P. (1962). *Therapist—Patient Expectancies in Psychotherapy*, Pergamon Press, New York.

Postman, L. (1951). 'Towards a general theory of cognition'. In J. H. Rohrer and M. Sherif (Ed.), *Social Psychology at the Cross-roads*, Harper, New York.

Storms, L. H., and Sigal, J. J. (1958). 'Eysenck's personality theory'. *Brit. J. med. Psychol.*, **31**, 228—246.

Yaktin, A., and Slater, P. (1965). 'Discontinuities and lumpiness in records of extinction'. *Brit. J. mat. and stat. Psychol.*, **18**, 69—86.

NAME INDEX

Note: Initials have been omitted except when confusion is liable to result. Numbers refer to pages in the text where the names are mentioned. For further information consult the lists of references.

SUBJECT INDEX